W0006778

A Creation of His Own:
Tappan's Detroit Observatory

Patricia S. Whitesell

Bentley Historical Library
The University of Michigan
Ann Arbor
1998

1935

Bentley Historical Library
The University of Michigan

ISBN 0-472-59006-5 (hb)
Copyright © 1998 — Patricia S. Whitesell

2001   2000   1999   1998   4   3   2   1

Written Inquiries:
Detroit Observatory
University of Michigan
1398 E. Ann Street
Ann Arbor, Michigan 48109-2051

Printed in the United States of America
on acid-free paper.

Cover and book design by Patricia S. Whitesell
Printed by University of Michigan Printing Services

Frontispiece	Jasper F. Cropsey's 1855 painting of the Detroit Observatory at the University of Michigan engraved by R. Leggett. (Bentley Historical Library)
Jacket Cover	Earliest known photograph of the Detroit Observatory, circa 1858. Although not confirmed, the man in the photo may be Franz Brünnow and the dog may be Henry Tappan's dog Leo. (Bentley Historical Library)
Jacket Back	Drawing of the 1857 Henry Fitz, Jr. refracting telescope at the University of Michigan. (Christopher Ray)
	Henry Philip Tappan (Bentley Historical Library)
Poem p. xii	Excerpt from a poem written by Henry Philip Tappan to Mrs. William W. Murphy for her scrapbook, October 5, 1873, Perry, p. 390.

To John

"I cannot speak of the Observatory without emotion.
No one will deny that it was a creation of my own."

Henry Philip Tappan

Preface and Acknowledgments

The preparation of this book has been a great pleasure. My research and writing, and my work directing the restoration of the Detroit Observatory at the University of Michigan, are challenges I have been privileged to undertake. These opportunities would not have been possible without the support and encouragement of Homer A. Neal, who as Vice President for Research and Interim President, with the support and backing of President and Mrs. Duderstadt, adopted the Detroit Observatory and placed his confidence in me to direct the restoration of the building, document its history, create a museum, and integrate the Observatory into the intellectual life of the University of Michigan. Frederick C. Neidhardt, as the subsequent Vice President for Research, added his backing to the project with his characteristic enthusiasm for science, and along with Provost Nancy Cantor effected my appointment as the first Director and Curator of the newly-restored facility. Francis X. Blouin, Director of the Bentley Historical Library at the University of Michigan, supported the publication of this book, thereby enabling a wide audience to share in the fascinating details of the observatory that was the creation of the University's first president, Henry Philip Tappan.

A Creation of His Own: Tappan's Detroit Observatory delves deeply into the Detroit Observatory's early biographical, architectural, and scientific history, beginning in 1852 with President Henry Philip Tappan's inauguration and tracing his efforts to implement progressive educational ideas on the Michigan frontier. Tappan's success in building the Observatory would result from a magical matching of scientific and applied interests: Tappan would have his scientific research laboratory, and the Friends of Science in Detroit would benefit from the applied uses of astronomical science. The book provides significant new, previously unpublished information and photographs that enrich the fascinating story of this physical legacy of the University's first President. It builds upon current literature about the Observatory by embellishing previously known factual details with extensive new information and illustrative quotes from primary sources, thereby threading together in one document the story of Tappan's Observatory. Over 100 photographs and illustrations, many never before published, bring the book to life.

For further reading, some excellent sources are:

- Charles M. Perry's biography *Henry Philip Tappan;*

- University of Michigan *Encyclopedic Survey* sections on the Tappan Administration (Vol. I), Astronomy (Vol. II), and Astronomical Observatories at Ann Arbor (Vols. II and III);
- W. J. Hussey's "General Account of the Observatory" in *Publications of the Astronomical Observatory;*

- Tappan's *Review by Rev. Dr. H. P. Tappan of His Connection with the University of Michigan,* and

- Howard Plotkin's "Henry Tappan, Franz Brünnow, and the Founding of the Ann Arbor School of Astronomers, 1852-63."

In addition, the extensive notes and bibliography at the end of this book offer numerous primary and secondary sources for further reading. Most important of all, the Bentley Historical Library of the University of Michigan holds a wealth of information and photographs related to the Detroit Observatory's fascinating history.

This research introduced me to an extensive network of individuals interested in the history of astronomy. Of this group, I express gratitude to Bob Ariail, Bart Fried, Bob Hambleton, Ken Launie, and other members of the Antique Telescope Society; Bob Bless, University of Wisconsin-Madison; Deborah Jean Warner, Smithsonian Institution; Nancy Langford, Dudley Observatory; Chris Ray, C. Ray Museum Studios; Frank Lorenz, Hamilton College; Steve Dick and Brenda Corbin, U.S. Naval Observatory; Ian Bartky; Owen Gingerich, Harvard University; the Greenwich Observatory; Jay Pasachoff, Williams College; John Williams, Albion College; Wulff Heintz, Swarthmore College; George Greenstein, Amherst College; Ian Elliott and Evart J. A. Meurs, Dunsink Observatory, Ireland; Rob van Gent, Boerhaave Museum for the History of the Natural Sciences, Leiden, The Netherlands; Kevin Johnson, Science Museum, London; Peter Hingley, Royal Astronomical Society, London; Henrik Glintborg, Tycho Brahe Planetarium, Copenhagen, Denmark; Hans Buhl, Steno Museum, Århus, Denmark; Ileana Chinnici, Palermo Observatory, Italy; Gerhard Hartl, Deutsches Museum, Munich; and Joerg Zaun, doctoral student at Technische Universität Berlin.

I am also grateful for assistance from Harvard University Archives; New York University Archives; Cooper Union Archives; Mary Lea Shane Archives of the Lick Observatory; Adelia Rasines, Newington-Cropsey Foundation; Archives Committee of the First Presbyterian Church of the City of New York (Dean Layton Hector); Leonard A. Walle, private photograph collection; Ralph Carmosino, New York City Parks Department; League for the Restoration and Preservation of Litchfield Villa; Metropolitan Museum of Art; New York State

Archives; Geoffrey Stein, New York State Museum; Frank Mason, Gurley Precision Instruments, Inc.; Cornell University Library; Schaffer Library, Union College; Cincinnati Historical Society Archives; Amherst College Library; Burton Historical Collection, Detroit; Michigan Historical Museum; Paul Gross, WDIV-TV, Detroit; Carol Mull, Kempf House Center for Local History, Ann Arbor; Quinn Evans/Architects, Ann Arbor; and Françoise Lambert, Conservatrice, Musée Historique du Vieux-Vevey, Switzerland.

Colleagues and students at the University of Michigan who assisted me include Homer Neal; Fred Neidhardt; Anne Duderstadt; Jim Duderstadt; Fran Blouin; Bob Warner; Anne Frantilla, Nancy Bartlett, Karen Jania, Marjorie Barritt, Malgosia Myc, Rachel Allen, Ken Scheffel, Ann Flowers, Jim Craven, and the entire staff of the Bentley Historical Library; Julie Truettner; Bill Hennessey; Patrick Seitzer; Freeman Miller; Dick Teske; Rudi Lindner; Gary Bernstein; Julie Ellison; Joanne Leonard; Bob Payne; Dennis Baker; Frank Marsik; Geoff Brown; Robin Meador-Woodruff; Susan Wineberg; Howard Markel; Ken Banas; Doug Covert; Dale Kocevski; Karen Woollams; Jackie Hoats; Connie Bridges; Lee Katterman; Mike Cook; Lori Mott; the Regents Office; Julie Doty, Ed Smith, Nancy Popp, Paula Yocum, and Doug Gowen of Printing Services; Colin Day and Ingrid Erickson of the University of Michigan Press; Rebecca Dunkle, Paul Barrow and many other staff members at the Hatcher Graduate Library; Special Collections Library; Buhr Facility; and Shapiro (Science) Library.

Many individuals contributed in diverse ways to the preservation of the Detroit Observatory during the 1970s through 1993. Their dedicated efforts were critical in preserving this important but often neglected structure. They are: Orren Mohler, Al Hiltner, Hazel Losh, John Hathaway, A. Craig Morrison, Dick Teske, Anne Cowley, Richard Frank, Wilbur Pierpont, Kingsbury Marzolf, Nick Steneck, Peg Steneck, Eagle Scout Ray Comiske, Fred Mayer, David Stockson, Paul Spradlin, Farris Womack, Eric Bram, historic preservation students Julie Truettner and Susan Wineberg, and architecture students Eric Murrell, Joel Zwier, and Tim Card. This list is partly generated from historical files rather than personal knowledge. If I have left out any contributors, which seems inevitable given the Detroit Observatory's many friends, the omission is not intentional.

Great numbers of people contributed to the physical restoration of the Detroit Observatory in 1997-98. I want to pay a very special tribute to the late David Evans, FAIA, of Quinn/Evans Architects—my friend, neighbor, and professional colleague who contributed so much to the restoration of the Detroit Observatory. Although he did not live to see the completion of the restoration work, he toured the Observatory shortly before his death, and his pride in the project was apparent. I greatly admired his creativity and scholarly orientation, his passion for history, his thoughtfulness, and his skill working with clients and other professionals. Regrettably, space limitations restrict me from recognizing everyone associated with the restoration project. The individuals who made management contributions included: Doug Hanna, Tom Abdelnour, Glenn Rittenger, Tom Schlaff, and Fred Mayer. Architects and contractors included: David Evans, Ilene Tyler, Scott Reinholt, and Mark Carter. Lloyd Baldwin served capably as the project's Field Representative and my assistant. Craftspersons included: Dennis Van Wagoner, Ron and Terry Isaac, Gary Ontko, Rick Kett, Fred Marion, Steve Hackbarth, Patrick Thornsberry, Phil Reed, Steve Correll, Ron Wilke, Kevin John, Kelly England, Jerry Milliken, Bill and Scott Miller, Bill and Harold Knapp, Bill Robin, Doug Lammers, Kevin Kooyers, Dave Dyer, Jamie Mayrand, Rick Luft, Ed Cantu, Jeff Jording, Brad House, Bob Epperson, Tom Bevins, Pat Jamison, Pat Monaghan, Ken Beaudry, Dave Gilbertson, Jerry Rutan, Ed Hollan, Steve Seebohm, and Dick Spigelmyer. Consultants included: Chris Ray, Frank Welsh, Malama Chock, Danielle Sheen, Diane Delatorre, Laurie Dickens, and Marcia Reed. I am grateful for and proud of the high quality of the restoration work.

Several members of my family assisted my research in their areas of knowledge and professional expertise, including my mother, Patricia S. Whitesell, a former university English instructor; my father, Don M. Whitesell, a retired organization planner and university instructor; my brothers, David R. Whitesell, a rare book librarian at Harvard University, and his twin, Steven M. Whitesell, a landscape architect with the New York City Parks Department; and my sister, Ann M. Whitesell, a photo archivist.

Most of all, I wish to thank my husband, John R. Wolfe, an environmental engineer and economist. As is the case with all of the projects I undertake, John provided constant encouragement, insight, and technical assistance, traveling with me to visit other observatories across the country and in Europe, and searching through archival materials—always eager to share in the details of the latest finding. It is to John that I have dedicated this book in recognition of our shared devotion to the Detroit Observatory.

Finally, I am grateful for the opportunity to explore the minds of such enormous personalities as Henry Philip Tappan and Franz Brünnow. Their impact on me personally, and on my research and scholarship, is complicated and profound. Tappan's scholarly writings are enlightening; Brünnow's hard work and exacting standards inspire me to set high personal standards, and his love of science sparks my curiosity to learn. Their letters and poetry let me into their minds, and their hearts. The discrimination they experienced here at Michigan is a lesson in openness and tolerance we must learn from and never forget. Tappan and Brünnow touched many, many minds and left an enduring legacy at the University of Michigan. This book locks into memory these special men: place it on your shelf, but think of them often.

University of Michigan   Patricia S. Whitesell
Ann Arbor, 1998

And yet as your book is a book of scraps,
This also a scrap must be,
Only a scrap of sense, perhaps,
Only a scrap of melody;
But most of all, a scrap of myself
Which I hope you will store on memory's shelf.

Henry Philip Tappan
October 5, 1873

Contents

List of Illustrations

Chapter 11   James Craig Watson

Chapter 12   Brünnow Departs for the Dudley Observatory

Chapter 13   Tappan and Brünnow Leave Michigan

Epilogue   Perpetuation of a Legacy

List of Appendices

A Creation of His Own:
Tappan's Detroit Observatory

Introduction

The year was 1852. In the previous year, a new constitution for the State of Michigan had instituted a change in the organization of the University of Michigan's Board of Regents. A new Board was now in place, and one of its first duties was to elect a President—the institution's first—to lead the University forward. "Let some distinguished man be placed at the head of the institution as chancellor, who can give it character and standing . . ."[1]

The appointment that followed would change the University of Michigan forever, creating an institution grounded in the usual classical course, but with an innovative new scientific course, and a University, or graduate, course that would guide Michigan toward its current status as a leading research university. Libraries, museums, and laboratories would be needed to create a scholarly environment of the highest order, and an astronomical observatory would be the first physical structure erected to facilitate the transformation. In this way, the University would create its first dedicated scientific laboratory.

The man, Henry Philip Tappan, whose vision and spirit of determination would lead the University to international prominence, would, reflecting back on his Michigan experience twelve years later, find his emotion fixed on the Detroit Observatory:

> I cannot speak of the Observatory without emotion.
> No one will deny that it was a creation of my own.[2]

The creation of the Detroit Observatory—the men and women involved in its success, and the science that took place there in the early years—is a fascinating, rich, and important part of the University of Michigan's history, the history of higher education, and the history of astronomy. This history reaches far beyond the University's original 40 acres: it extends across the country—and across the ocean—into the lives of the world's most prominent scientists and the nation's business leaders.

The Detroit Observatory, so named by Tappan to honor the city of its major benefactors, was to become prominent in both the academic and scientific life of the University and in the applied world of business and commerce. Tappan was exceedingly proud that his observatory would stand out from all others through the exceptional quality of its instrumentation, and that he built it at a low cost[3] overall:

> The Observatory, one of the first[4] in the world, has been erected and furnished at less cost than any other of the same rank on record.[5]

And, the Observatory, in a sense, became part of Tappan's family through the recruitment of its first director, Franz Brünnow, who became the husband of Tappan's only daughter, Rebecca. Tappan later commented that "in some sort, I became thus wedded to the Observatory."[6]

Figure 1.1. Henry Philip Tappan. A larger than life-size portrait by Alvah Bradish, who painted many distinguished figures in Michigan history, and served as an instructor of Fine Arts at the University of Michigan. The plaster medallion of Tappan on the wall is the image of a real version Bradish created. (Bentley Historical Library)

Chapter 1

Henry Philip Tappan

In order to fully understand the path that led to the creation of the Detroit Observatory, one must come to know the man who created it, and the influences on his life that prompted him to establish an observatory as a priority of his presidency at Michigan. This man was Henry Philip Tappan, the philosopher and scholar in whom the Regents of the University placed their trust and confidence, inaugurating him as the first president of the University in 1852. [Figure 1.1]

Tappan had not been the Regents' first choice for the position. General Lewis Cass and others had nominated Professor Charles Anthon, a distinguished classical scholar from Columbia College. The Governor of Indiana had recommended Dr. Elizur Deming of Lafayette, Indiana. When it came to the first vote in June 1852, the Honorable Henry Barnard, a distinguished educator and founder of *The American Journal of Education,* was unanimously elected. When Barnard declined, the offer was extended to the Reverend William Adams, a Presbyterian pastor from New York City, who in 1872 accepted the Presidency of Union Theological Seminary. Adams declined. The third vote was split with three votes going for John H. Lathrop, Chancellor of the University of Wisconsin, and five votes cast for Henry Philip Tappan. The three Regents who supported Lathrop changed their votes, and Tappan was unanimously elected on August 12, 1852.[7]

The University had engaged in discussions with Tappan as early as March of 1852. George Bancroft, the historian and former secretary of war, had been approached by the Regents, but he declined to be considered, offering Tappan's name as a possible candidate. Regent Charles Palmer, who had been a student of Eliphalet Nott's at Union College, Class of 1837, had written to Tappan to enquire about his interest in the presidency. Tappan responded, indicating that he had lived in New York so long, and had established such strong connections there that, as a matter of course, he declined offers that required relocation. Nonetheless, the position interested him because it related to his favorite subject: the perfection of the system of public higher education. Tappan concluded that he was "indifferent" but was willing to enter into correspondence to learn

Figure 1.2. Jasper F. Cropsey's 1855 sketch of the President's House in Ann Arbor. (Bentley Historical Library)

more about the specific details of the position.[8] It was after much correspondence and a visit to Ann Arbor in late September that Tappan accepted the appointment. The Tappans—Henry, Julia, and their children, John and Rebecca—moved to Ann Arbor in October 1852. [Figure 1.2]

Tappan was born in 1805 at Rhinebeck on the Hudson River, north of New York City.[9] His ancestors had emigrated from Lorraine to Holland, and then to America. Tappan's father, Peter, was an officer in the Revolutionary army, serving with distinction at Yorktown when Cornwallis surrendered. Henry's grandfather, Christopher Tappan, was active in the politics of Kingston and Ulster counties. When British troops invaded New York City, he had sacrificed his home and possessions in order to rescue the State's documents. The Tappan family had intermarried with the Clintons, and his mother, Ann DeWitt, was directly related to the famous DeWitts of Holland. General George Clinton, the first and seven-time governor of New York State, who was married to Tappan's great-aunt Cornelia Tappan, was Vice President of the United States when Henry was born. DeWitt Clinton, governor of New York State and canal commissioner for the Erie Canal project, was Tappan's third cousin.

Although Tappan's family was prosperous, and held education as a high value, Tappan's father lost his savings to unfortunate endorsements after the War of 1812, and Tappan was left to his own resources. After attending district school and the Greenville Academy, he withdrew at the age of fourteen to teach for two years at the Dutch Reformed Church of New Paltz in order to earn money sufficient to enter Union College in Schenectady, New York, where his brother, Charles DeWitt Tappan, was enrolled. Henry entered Union College in 1822 as a sophomore, apparently having convinced officials that he was worthy of advanced placement. He was admitted to Phi Beta Kappa the following year. Although he continually struggled to make his college payments, Tappan arranged loans with Professor Andrew Yates, repaying him in full following his graduation in 1825. Tappan was awarded advanced standing for his subsequent studies at the Auburn Theological Seminary, primarily because he had elected Hebrew in his final term at Union College, which reduced his theological studies by a full year.[10] Following his graduation in 1827, he began his ministerial career as assistant pastor in the Dutch Reformed Church at Schenectady.

Tappan was a favorite student of Eliphalet Nott, the president of Union College. [Figure 1.3] It is notable that by 1845, thirty of Eliphalet Nott's Union College graduates had become presidents of American colleges.[11] Nott's influence on Tappan is quite apparent. Tappan's inaugural speech, made on December 21, 1852 in Ann Arbor, highlights many of the ideals Eliphalet Nott

Figure 1.3. Eliphalet Nott of Union College. Tappan was Nott's protégé. Painting by Henry Inman, 1839. (Union College, Schenectady, New York)

prescribed for Union College. Nott had been the first to implement a Parallel Scientific Course at an American college or university. Thomas Jefferson had introduced ideas for academic reform that were influenced by European models, but the idea of introducing a scientific course that would venture beyond the classical course was brought to this country by George Ticknor of Harvard University, who had been educated in Göttingen, Germany. However, Ticknor's ideas, which were eloquently written and well-publicized, were rebuffed by a Harvard that wasn't prepared for such change. In contrast, Nott, whose Union College at that time was comparable in stature with Harvard and Yale, introduced his scientific course quietly and successfully in 1828.[12] Nott's Parallel Scientific Course was soon elected by one-third of the students at Union College. It was centered around the professions of engineering, law, medicine, mining, and scientific subjects that employed mathematics, such as surveying and astronomy, a model Tappan was to follow at Michigan.

In 1845, Nott's plans called for an observatory to be built on the ridge high above and to the south of the Union College campus. Nott had been awed by the solar eclipse of 1806, an event he referred to as "one of the most sublime and awful spectacles this age has witnessed."[13] In anticipation of the event, Nott had a telescope in place, and he recorded the eclipse with scientific exact-

ness. Elsewhere in New York State, James Fenimore Cooper, then a boy of 17 years, observed the "appalling eclipse [that] swept in from the east and deepened into darkness at noon"[14] through a piece of smoked glass, and he recorded the drama of the event as perceived by the non-scientist: "Many of the inhabitants here were seized with horror, some broke forth in supplications, some fainted and some were flung into convulsions."[15] The widely held view of this dramatic event was that it was "designed by the creator to reveal his Omnipotence to our race."[16] [Figure 1.4[17]]

Astronomy was a science that held both awe and fear at that time. Nott himself had once held the view that "the earth would be destroyed in a cataclysm of converging sun and planets and shattering meteors,"[18] but he resolved

Figure 1.4. Solar eclipse of 1869 at Burlington, Iowa, taken by photographers who accompanied James Craig Watson and other astronomers (see Figure 11.3). (*Morton Memorial*, 1905)

these fears by adopting a more hopeful view of communication with other beings that would be made possible only through the pursuit of astronomy. He envisioned a time when "our improved instruments shall enable us to recognize their signal, and to give back by telegraph from our sidereal watch-towers the signs of recognition."[19]

Just as Nott was highly influential in forming Tappan's view of the value of a scientific curriculum, Tappan's marriage influenced, and may have assisted, Tappan's rise in higher education. Education, at that time, was largely (though not exclusively) influenced by and accessible to only the wealthy. In 1828, Tappan married Julia Livingston of Skaneateles, New York, the eldest daughter of Colonel John W. Livingston, Jr., the United States Marshall of the Northern District of New York. [see Figure 13.4] The Livingstons were a large and promi-nent family of great wealth and status in New York State. Tappan's charm, intellect, scholarship, and family ties, no doubt, placed him in proximity with, and in the good graces of, the Livingston family prior to his marriage, and the marriage helped to place him in an elite circle of New York's most prominent and influential citizens. The marriage ceremony took place on April 17, 1828 by the Reverend Dr. Perrine of the Auburn Theological Seminary, where Tappan had studied.[20]

The Livingstons were a family of American statesmen, diplomats, and jurists. Through marriages, the family established connections with other prominent New York families such as the Van Rensselaers, Schuylers, Montgomerys, and Armstrongs. As an example of the family's prominence, Robert R. Livingston was the law partner of John Jay, American statesman and the first Chief Justice of the United States. He was also a delegate to the second Continental Congress and a member of the committee to draft the Declaration of Independence. Later, Thomas Jefferson appointed him as minister to France, where he conducted the negotiations that resulted in the Louisiana Purchase. The Livingston family's mansion, called Livingston Manor, was located north of New York City along the Hudson River, not far from Rhinebeck. Charles DeWitt, Tappan's maternal grandfather, had managed the estate, later com-pleting his career in politics, including membership in the Congress of the Confederation at Annapolis. It is thought that Tappan was named after Henry Brockholst Livingston (1757-1823), who served in the American Revolution and became a lawyer and ardent Jeffersonian, eventually receiving appointment by Jefferson to the United States Supreme Court; and Philip Livingston (1716-78), a successful merchant who signed the Declaration of Independence and later helped to found King's College (Columbia University) and the New York Society Library.[21]

Tappan took a position at age twenty-three at Pittsfield, Massachusetts[22] as minister of the First Church of Christ, Congregational, but left in 1831 after three years in order to recuperate in the West Indies from a bronchial infection. His health improved quickly, and he returned to New York in 1832 where he was called to the newly established University in the City of New York (now known as New York University and hereafter referred to as NYU) as the chair of moral and intellectual philosophy, remaining there until 1838 when he was dismissed. The faculty of NYU had in 1833, with the exception of Tappan, appealed for the removal of Chancellor Matthews on the grounds that he interfered with faculty attempts at discipline in favor of avoiding the alienation of parents who might be potential donors to the institution's endowment. But, by 1835, Tappan's loyalty to and respect for Matthews had slipped, and Tappan became a faculty leader in suggesting that the Chancellor employed dictatorial methods and mismanaged institutional funds. The University's Council, which was also losing confidence in Matthews, undertook an investigation which confirmed financial mismanagement. As a result, so many Council members resigned that the purged Council was able to discipline the disloyal by discharging seven faculty members, Tappan included.[23]

Tappan was briefly the head of a private seminary for young ladies located in Leroy Place, 113 Bleeker Street, New York.[24] At this time, Tappan threw himself into the study of philosophy, publishing several important works including his *Review of "Edwards's Inquiry into the Freedom of the Will"* (1839); *Doctrine of the Will Determined by an Appeal to Consciousness* (1840); *Doctrine of the Will Applied to Moral Agency and Responsibility* (1841). He added to this list the *Elements of Logic* (1844); *University Education* (1851); and *A Step from the New World to the Old, and Back Again; with Thoughts on the Good and Evil in Both* (1852). His works, and his philosophy of education, were largely influenced by the eclectic French educational leader and philosopher, M. Victor Cousin, who said of Tappan's *Elements of Logic*, "It is equal to any work on this subject that has appeared in Europe."[25] [Figure 1.5] George Bancroft, who had suggested Tappan as a candidate for President of the University of Michigan after declining the offer himself, was a devoted believer in Cousin's philosophy. Rooted in German idealism, Cousin's early metaphysical works greatly influenced American thought. Later in his career, he was attacked by clergy for his defense of the liberty of science.[26]

Figure 1.5. French philosopher M. Victor Cousin was Tappan's mentor. (Pictorial History of Philosophy, 1959)

In 1845, Tappan received from Union College the honorary degree of Doctor of Divinity on the occasion of Union's 50th anniversary, and in 1854, he received from Columbia University an honorary LL.D. Three years later, he was invited again to Union College to deliver the principle oration on the occasion of the 50th anniversary of the Philomathean society, a "secret" society to which Tappan had been admitted in his first year of study. Tappan was honored by Philomathean as First Orator in his senior year, which permitted him to read one of his poems at commencement. He joined the American Association for the Advancement of Education in 1855 and was immediately elected president, presiding over the group's annual convention in Detroit in 1856, and in Albany the following year. What was perhaps his most significant academic honor was bestowed upon him in 1856 with his election as a corresponding member of the Institute of France, no doubt championed by his mentor M. Victor Cousin. In 1858, he returned again to Union College to make an address when the cornerstone was laid for the new 16-sided architectural marvel constructed in the center of campus—Alumni Hall, which was later named the Nott Memorial.[27]

The Tappans had five children, though only two lived to adulthood. Infant mortality was, unfortunately, quite common, and the Tappans faced tragic loss in this way three times. Their first child, Caroline, was born on November 11, 1829; she died March 6, 1831. John Livingston was born August 5, 1831, and he became a very capable librarian at the University of Michigan. [Figure 1.6] His life ended unexpectedly at the age of forty-one, in January 1872, on his way home from a stay in Paris. John's birth was followed by Anna Lydia on November 22, 1832; she died in infancy on August 29, 1833. Rebecca Lloyd, nicknamed "Barbie," was born May 27, 1836, the only child who lived a full life; she died July 6, 1893. Anne DeWitt was born May 22, 1838, but died December 7, 1843. Tappan, who often composed poetry, wrote the following lines in the Tappan family *Bible* below the infant death entries:

Figure 1.6. John Livingston Tappan served as Librarian at the University of Michigan. (Bentley Historical Library)

> Oh! when a Mother meets on high
> The Babes she lost in infancy,
> Hath she not then, for pains and fears,
> The day of woe, the watchful night,
> For all her sorrow, all her tears,
> An overpayment of delight?[28]

By the time Tappan was inaugurated as President of the University of Michigan in 1852, he had become internationally recognized as a philosopher and educator. He was offered but declined the opportunity to return to his former chair at NYU in favor of the presidency at Michigan. Tappan saw in Michigan the opportunity to "change the wilderness into fruitful fields"[29] by putting into action the ideas he had formulated regarding university education, modeled after the Prussian system:

> . . . the institutions of Prussia, like ancient learning and art, stand before us as models which we are constrained to admire, to approve, and to copy. The institutions of Michigan are yet in their infancy, but we think there is promise of a bright career, of a full and ripe development, which cannot well disappoint us.[30]

Tappan's plan, as influenced by Eliphalet Nott and M. Victor Cousin, and by Tappan's travels in Europe, was to integrate the scholarly, classical course of study with a basic and applied scientific course. In Tappan's view, the time had arrived for the United States to establish its own scholars rather than depending upon European intellectuals. Further, he believed that our nation needed to intertwine the literary, scientific, and artistic life with the political and commercial. This combining of scholarly study of the classics with basic and applied science was to set the model for the American research university.

Henry Tappan saw a world with two commanding ideas: utility and beauty. In his view, science and scientific research directly benefited the applied needs of our society. In his inaugural speech, he said:

> The world is useful: the world is beautiful. But the knowledge we have gained by experience enables us, under the light of these ideas, to conceive of other forms of utility and beauty besides those which are presented in the rude uncultivated nature around us. Then comes the application of our skill and industry . . . The activity of human thought—scientific cultivation has done it all.[31]

Tappan's view of higher education was not new to Michigan. Between the years 1835 and 1837, when the State constitution was being formed, General Isaac E. Crary, Michigan's first representative in Congress and delegate of the approaching Convention, and the Reverend John D. Pierce, a missionary, were largely responsible for shaping the educational plan for the State. M. Victor Cousin's report on the Prussian educational system, entitled *Rapport sur l'état de l'instruction publique en Prusse*, had impressed them. When General Crary was made the chairman of the committee on education, he carried forward Cousin's ideas.[32] This prior familiarity with the integrated curriculum may have aided its acceptance at Michigan to some extent.

Personally, Tappan had a commanding presence, and was widely respected—even awed:

> What a man: He had that indefinable thing called presence. Tall, erect, dignified, commanding, majestic, you might have thought him Jove with his hands full of thunderbolts. A finely chiseled face, hair that always seemed tossed by the wind. He was of the classic mould—a figure from the antique world. He had the pose of the great statues, the pride and bearing of the intellectual Greek, of the conquering Roman, and he stood in the wide free air as though within his veins there flowed the blood of a hundred kings. . . . He had breadth and scope, resource, learning, he was original, thoughtful and profound and above all he had a sense of justice and bravely spoke his thought. He spoke as the thundercloud speaks. He was an immense personality.[33]

The University of Michigan was about to embark upon an exciting period of its history, led by one of the most dynamic, visionary leaders in the history of higher education—and an observatory was one of the first orders of business.

Figure 2.1. Astronomer Paul A. Smith using a sextant and artificial horizon on the roof of Detroit Observatory in 1922, with the recently built "Old Main" Hospital visible in background. (Bentley Historical Library)

Chapter 2

The Emergence of American Astronomical Science

Astronomical science in this country can be traced back to our nation's earliest beginnings. Knowledge of astronomy was brought to America by the first explorers (because astronomical methods are essential to navigation), and later by early settlers. Astronomy had great utility to the early explorers who surveyed this country. Meriwether Lewis, on the 1803 Lewis and Clark expedition through the Louisiana Territory to the Pacific coast, drew upon what he had learned in Philadelphia about the use of celestial observation to determine position.

Determination of latitude was performed using a sextant or octant to determine the altitude of the sun at high noon, thereby establishing the distance from the equator. [Figure 2.1] Determination of longitude, however, required a chronometer, an instrument designed to keep time with great accuracy, that was set to Greenwich mean time prior to departure into the field. Longitude could be determined by timing the passage of a star directly overhead. Although Lewis had purchased the best chronometer available, timekeepers were not yet sophisticated enough to function reliably, and the rigors, the elements, and the temperature extremes to which Lewis' chronometer was subjected during the journey, no doubt, caused inaccuracies. Though Lewis was accomplished at identifying stars, having benefited from Thomas Jefferson's interest in and knowledge of astronomy,[34] he was forced to estimate longitude from the position of the moon relative to the sun and particular stars, a less accurate alternative to use of a chronometer.

In 1812, when a grid pattern was laid out for New York City's streets, a large section of land, called Observatory Place, was designated as the future site of an observatory. [Figure 2.2] An observatory was never established at this location, but this evidence of planning for an observatory at this early date in the City's history speaks to the importance of astronomy at the time. Around 1815, Harvard College considered the establishment of an observatory, but committees assigned to review the idea found the cost of such an endeavor to be prohibitive.

John Quincy Adams later suggested in 1823 that an observatory be established at Harvard, but there was great opposition. Two years later, Adams

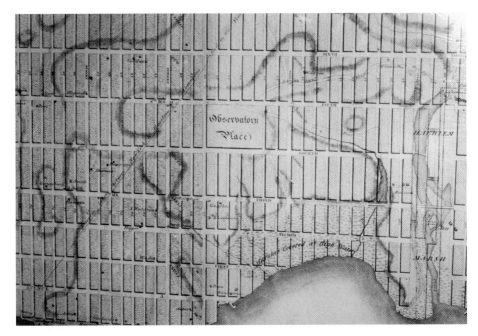

Figure 2.2. Observatory Place was set aside as future site for an observatory in 1812 when New York City was laid out in grid form. (Patricia S. Whitesell, taken at New York State Museum, Albany)

proposed the establishment of a national observatory, with the objective being to "increase knowledge by new discovery."[35] In Adams' view, "there is not one study in the whole circle of the sciences more useful to the race of man upon earth, or more suited to the dignity of his destination as a being endowed with reason, and born to immortality, than the science of the stars . . . The history of Astronomy has been, in all ages, the history of Genius and Industry, in their blazing light and untiring toil, patronized by power."[36]

Astronomy was considered to be the science of the elite in the early days. [see Figure 3.4[37]] Many prominent men, and some women, owned telescopes and pursued astronomy primarily as a hobby. Adams' interest in astronomy manifested itself as a personal resolve to remove the control of time from nature, as determined by the sundial and the almanac, and place it under the control of educated scientists and the government. But, his idea was ridiculed, and his phrase "light-houses of the skies" was used to mock his efforts to establish a national observatory, because the Coast Survey was the priority at that time. Adams cautioned against reliance on European observatories, which numbered in excess of 130. Adams said ". . . are we not cutting ourselves off from the means of returning light for light, while we have neither observatory nor observer upon our half of the globe, and the earth revolves in perpetual darkness to our unsearching eyes?"[38]

Nearly two decades passed before Adams' vision was achieved. Small steps had helped to clear the way for observatories to be established. One early step was the creation of a small observatory known as the Depôt, which was quietly set up by Lieutenant Wilkes of the Coast Survey near Capitol Hill in Washington in the early 1830s, and used primarily to rate chronometers for the Navy's fleet of ships. In 1834, the publication of John Herschel's *Treatise on Astronomy* introduced advances in theoretical astronomy—the dynamics of the universe, and the orbits of the planets based on gravitational forces—that had great influence on public attitude toward astronomy. In 1838, Elias Loomis established the Hudson Observatory at Hudson, Ohio, and in Philadelphia, Sears C. Walker opened the Philadelphia High School Observatory. West Point Observatory was established in 1839, followed in 1844 by a National (Naval) Observatory and Georgetown Observatory at Washington, D.C. The Cincinnati Observatory opened in 1843, the result of tireless effort by its director, Ormsby M. Mitchel. At Mitchel's invitation, John Quincy Adams, despite his advancing age and fragile health, made the long journey to Cincinnati to lay the cornerstone at the building's dedication. He would not miss this opportunity to celebrate the "perseverance to accomplish what [he had] promised."[39]

Explorations of the South Seas, which were launched in 1838, re-focused attention on the need for astronomical observations to determine the longitude, an essential element in keeping ships on course. The heavens cooperated, too, by captivating public attention with spectacular displays of meteor showers and awesome comets—their long tails blazing across the sky. [Figure 2.3]

Figure 2.3. Comet of 1843 as illustrated in James C. Watson's *A Popular Treatise on Comets*, 1860.

Comets were heralded by the educated as opportunities to expand knowledge, but others were superstitious that they might be "forerunner(s) of pestilence, famine, or some other dreadful calamity."[40] Gradually, astronomy became more accessible by the general public, and with this change came a wider acceptance of astronomical science.[41]

Colleges and universities recognized the value of astronomy much earlier than a national acceptance was achieved. It was around 1828 that astronomical observatories began to appear on campuses in the United States. Yale College located a telescope in the steeple of one of its buildings in that year. Williams College, in Williamstown, Massachusetts, began construction of an observatory [Figure 2.4] in 1836. This observatory still exists today, though it has been moved short distances several times. It is the oldest extant observatory in America. The building was named the Hopkins Observatory after astronomy professor Albert Hopkins, brother of the College's President, who cared so deeply about establishing an observatory that he assisted the masons and other tradesmen by contributing his own labor toward the construction of the small, stone building, with additional assistance provided by students. President Hopkins made it known at the dedication of the Observatory that his interest in its establishment was to counter the prevailing movement in higher education toward the practical in favor of elevating students' thoughts "toward that fathomless fountain and author of being, who has constituted matter and all its

Figure 2.4. Hopkins Observatory at Williams College, built in 1836, is the oldest extant observatory in America. (Detroit Observatory Collection)

Figure 2.5. Amherst College Observatory, built in 1847, was designed after alumnus Orson S. Fowler's invention of the octagon-shaped building. Note the building's unusual faux stone stucco finish (see Chapter 9). (Amherst College Archives and Special Collections)

accidents as lively emblems of the immaterial kingdom."[42] The original telescope, a Herschelian reflector, is no longer extant, having been traded sometime prior to 1852 "for the bones of some animal found in Pennsylvania."[43]

Astronomy's popularity had been slow to advance in America's first half century, largely due to a failure to advance beyond what was called practical astronomy—the measurement of celestial positions and the determination of local time and geographical coordinates. Advancements were needed, and the heavens cooperated. It was the comet of 1843 [see Figure 2.3] that provided the final impetus needed to create a new observatory at Harvard College. With this facility and a 15-inch German-made (Merz and Mahler) refracting telescope, Harvard College took the lead in training and instruction in observational astronomy.

Between 1845 and 1854, several observatories were established including Sharon Observatory, near Philadelphia; Tuscaloosa Observatory in Alabama; several private observatories at New York residences, including those of amateur astronomer Lewis Rutherfurd and professor Richard H. Bull of New York University; Friend's Observatory in Philadelphia; Amherst College Observatory[44] in Massachusetts; [Figure 2.5] Charleston Observatory in South Carolina; Dartmouth College Observatory in New Hampshire; [Figure 2.6] Mr. Van Arsdale's private observatory in Newark, New Jersey; Shelby College Observatory, mounted atop a campus building in Shelbyville, Kentucky;

Figure 2.6. Dartmouth College Observatory, erected in 1853, was designed by Ammi Young. (Collection of Leonard A. Walle)

Mr. Van Duzee's private observatory in Buffalo, New York; Mr. Campbell's observatory atop his home near NYU; and the Hamilton College Observatory (later named the Litchfield Observatory[45]) in Clinton, New York.[46] [see Figure 8.6]

By 1854, when Tappan's Detroit Observatory was completed, there were about 25 observatories in this country. Most observatories were in the East, in states such as Massachusetts, Pennsylvania, and New York, and in Washington, D.C. Tappan's was the first observatory established in Michigan, and second only to the Cincinnati Observatory (1843) in the Midwest. Another 25 years would pass before observatories would be established further to the west, such as the Washburn Observatory (1879) in Madison, Wisconsin. It wasn't until the 1880s that Michigan gained a second observatory, at Albion College.

During the 1840s and 1850s, the academic observatories worked closely with the government surveys, such as the national Coast Survey and the Lake Survey, on endeavors involving practical astronomy. It was the establishment of astronomical observatories across the country that enabled surveyors to have access to precise astronomical time by which they could accurately set their chronometers as they headed out to conduct land and coastal surveys. The United States Coast Survey had been formed in 1807 by Thomas Jefferson to chart the nation's coasts and land formations for the safe navigation of ships. On average, three hundred vessels were annually wrecked along the Atlantic Coast.[47] While the determination of latitude for travel from north to south was relatively easy, travel from east to west required precise longitude calculations, and accurate time determined by astronomical techniques was necessary to determine longitude.[48] The War of 1812 and other factors caused delays that postponed surveying efforts until 1832. The Coast Survey made numerous contributions to science, including studies of magnetism and sea currents. Many factors, including their expertise in predicting variations of the compass, led to the subsequent involvement of the Coast Survey in land surveys.

As the country was settled, and as railroads and canals were established across the country, the need for accurate land surveys became more critical. Surveyors needed accurate coordinates to use as a basis for their surveys. Public surveys of the State of Michigan began in 1815 and were completed in 1857. A principal meridian line was established running due north from the mouth of the Auglaize River, a branch of the Maumee River in Ohio, and a base line was located to intersect the meridian line along the northern boundary of Wayne County, running due west across the State as the northern boundary of a line of other counties, across to Lake Michigan. These coordinates served as the points of reference for all future surveys.

Early surveys were grossly inaccurate, in some cases misrepresenting the position of cities by three miles or more, owing to the imperfection of the instruments used, the presence of iron deposits that interfered with compass readings, and the "ignorance, carelessness and dishonesty" of some survey-ors.[49] The solar compass, invented by William A. Burt[50] and patented on February 25, 1835, helped to alleviate the interference of iron deposits on the accuracy of the compass needle. [Figure 2.7] The solar compass, through the ingenious use of the sun's rays to determine the exact position of a north-south line, made it possible to conduct accurate surveys in areas where the presence of minerals made the conventional compass useless. Geologist Bela Hubbard said of his use of the solar compass for the Upper Peninsula surveys in 1845:

> This accuracy has been attained by the exclusive use, by all the parties, of "Burt's Solar Compass," an instrument too well known to need more than a bare

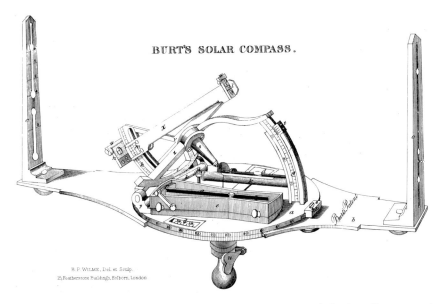

BURT'S SOLAR COMPASS.

B. P. WILME, Del. et Sculp.
15, Featherstone Buildings, Holborn, London

Figure 2.7. Burt's Solar Compass, invented by William A. Burt in 1835, helped to alleviate interfer-ence from iron deposits on the accuracy of the compass needle. (William A. Burt, 1844)

allusion, but the great value of which has been more than fully confirmed during the surveys of the past season. . . . It seems difficult to imagine how the lands could have been run with the ordinary compass.[51]

The invention was commended by Sir John Herschel in 1851 after it had been perfected the previous year, and it won Burt the medal of the Franklin Institute and a prize medal at the London World's Fair.

Precision land surveying was very important to the railroad in its desire to acquire land to lay track for the railroad. If not properly surveyed, property disputes could subsequently arise that would have serious consequences if tracks had to be moved. Accurate surveys were equally important to the acquisition of land for mineral explorations. Boundaries were set by surveyors' marks blazed on trees with various cuts, letters, and numbers to signify a north-south line, distances between points, and so on. [Figure 2.8] Where trees were not present, mounds of earth or stone were placed around a wooden post.

As the railroads expanded, time became a very important factor in successful and safe railroad operation. The industrial revolution brought with it the post office, railroads, and the telegraph, all of which brought into conflict the differences in local time. There was great confusion because each city, town and village had its own local time. A person planning to catch an outgoing train or meet an incoming visitor at the railroad station needed to know what time to arrive; towns needed to predict when the mail would arrive, and depart; and

Figure 2.8. Surveying class at the University of Michigan circa 1875, posing with surveying instruments, axes, and chains. (Bentley Historical Library)

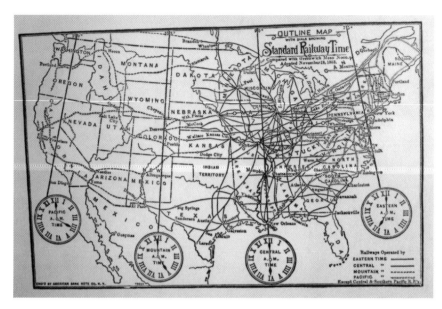

Figure 2.9. Standard railway time zones were not established until 1884. (Patricia S. Whitesell, taken at Greenwich Observatory)

railroads with many cars sharing the same track needed to coordinate two-way traffic in order to avoid head-on collisions (which were an unfortunate, though not uncommon, occurrence). Accurate time was essential:

> *Time* exactly obtained and kept is the regulator of this complex system of moving parts, which, in theory at least, should resemble one great piece of clock-work. To make things "move like clock-work" is not merely a figure of speech, then, here, where our lives depend on the accuracy of the conductor's watch . . .[52]

Surprisingly, it was not until 1884 that a uniform system of time was finally established. [Figure 2.9] Prior to 1884, every city, town and village kept its own time, usually based on inaccurate calculations of "sun time." The *Old Farmer's Almanac* made widely available a guide to using the stars to determine time. Such methods were imprecise, but they served adequately until advances in transportation made possible rapid travel across the country. Problems related to time then began to escalate, keeping pace with the increase in mobility.

Another need for accurate time that is, perhaps, less obvious, relates to "the security of commerce and the punctuality of business."[53] Accurate time was particularly important to business endeavors, and especially to bankers whose transactions ended at a designated time each day. Differences in time could result in serious disputes, disappointments, and disastrous financial losses.

Observatories provided accurate time, and also proved useful as scientific stations for the pursuit of meteorological science. The importance of weather forecasting to astronomical observers is evident. The prediction of weather conditions also held advantages to other citizens that were valued in ways similar to the calculation of time:

> If we could only foretell the weather a few days, or even a few hours in advance, it would be a boon of the greatest value to agriculture, navigation, and indeed to all the arts of life.[54]

Accurate prediction of the weather could save the farmer from planting his crop the day before a heavy rain that would wash away his seed, or save the sailor from shipwreck by a devastating hurricane. Of less consequence, yet important to the average citizen, a weather forecast could save the disappointment of a soggy picnic or a windy buggy ride.

The Detroit Observatory kept weather records from its beginning. Initially, data were kept to track temperature variations that could adversely influence the performance of the meridian circle telescope. Then, in October 1854, a complete set of meteorological instruments was obtained, and data for the documentation and prediction of weather were kept by Professor Alexander Winchell, who sent them on to the Smithsonian Institution. A windvane[55] was installed on the roof of the Observatory's west wing, and an anemometer captured wind speed on the roof of the director's residence. A wooden box with louvers to permit air circulation, called a weather shelter, was attached to the front of the west wing of the building. It protected the weather instruments that hung inside the box. A catwalk was constructed in 1908 between the original building and the director's residence to facilitate access to the weather instruments and to portable telescopes and other instruments used on the rooftops; another was added later to permit access to the roof of the 1908 addition. [see Figures 9.7, 9.13, 10.3, 10.7, and 10.9]

When Mark W. Harrington was appointed director of the Observatory in 1879, his keen interest in meteorology greatly expanded the Observatory's related activities and increased the importance of Michigan's role in the national meteorological community. John M. Schaeberle, Harrington's assistant, conducted the bulk of the astronomical work. Meteorological records were taken three times a day, recorded in ledgers,[56] and sent to the State Board of Health at Lansing. In 1884, Harrington founded the *American Meteorological Journal*, which was the first journal of its type, and served as editor until 1892. His expertise and national reputation led to Harrington's appointment in 1891 as the inaugural director of the United States Weather Bureau. William J. Hussey became the director of the Detroit Observatory in 1905, and upon his recommendation in 1907, a United States Weather Bureau station was established at the Observatory.

Chapter 3

Tappan's Observatory:
The Detroit Donors

The Detroit Observatory was completed in 1854. A priority of Tappan's agenda for instituting the scientific course at the University had been the establishment of an astronomical observatory. In fact, astronomy was central to his thinking about the establishment of a scientific course at Michigan. Time and space were at the center of Tappan's educational solar system. In his view, the great planetary masses (including the earth) that defined time and space composed the universe. Tappan considered this to be the essence of astronomy. To him, all of the sciences—physics, geometry, mineralogy, geology, chemistry, and even biology, physiology, zoology, and botany—were historically connected with astronomy.[57]

A building would be required to house the scientific apparatus of an observatory. Tappan's firm belief was that university buildings ought to be reserved for "the great things [letting] the small things take care of themselves."[58] Tappan's plan called for new campus buildings to house an observatory, library, museum, laboratories, scientific apparatus, and eminent professors. Students were to live in local rooming houses rather than dormitories in order to reserve buildings for instruction, thereby integrating students into the community in order to temper their potential for incivility.

Astronomy and other areas of science and mathematics had been taught at Michigan prior to Tappan's arrival, but no equipment had been purchased. The earliest record of a desire for astronomical instruments was in 1849 when the Regents expressed regret at not having any, though they pointed out that a law permitted the Board to purchase apparatus.[59] In 1850, the curriculum included scientifically useful courses such as algebra and geometry for freshmen; plane and spherical trigonometry and analytical geometry and calculus for sophomores; natural philosophy, chemistry, and mineralogy for juniors, with astronomy in the third term (taught using Olmsted's text[60]); and geology in the senior year. Professor George P. Williams taught natural philosophy and mathematics, including astronomy; Abram Sager taught zoology and botany; and Silas Douglass taught chemistry, mineralogy, and geology.

Figure 3.1. Henry Nelson Walker portrait by Alvah Bradish. Walker championed fundraising for the Detroit Observatory and contributed funds to cover the entire cost of a meridian circle telescope of the finest quality. (Burton Historical Collection)

Tappan believed that the University belonged to the people, and it is to the people that he turned for help in financing the construction of the physical campus, and an observatory in particular. Henry N. Walker [Figure 3.1] of Detroit was present the day Tappan delivered his inaugural address. At the conclusion of Tappan's speech, Walker stepped forward to offer his assistance in raising funds to construct an observatory. Obviously, Walker, who had significant interests in the railroad business, had identified with Tappan's position that scientific endeavors have both a scholarly and an applied value. Tappan had said:

> . . . can we truly be called a nation, if we cannot possess within ourselves the sources of a literary, scientific, and artistic life as well as of a political and commercial?[61]

Perhaps Walker took particular note when Tappan said:

> We aim not merely to equal, but even to surpass the old nations of the world, in our manufactures, our steamboats, and our railroads. We level the forests in a day, lay down our tracks, and startle the old world with the sounds of our engines. . . . substantial railroads in every direction . . . lead the astonished traveler through villages, towns and cities, which have sprung up by the magic of that industry whose divine mission it is to change the wilderness into fruitful fields . . . Let us make men as well as houses and railroads. Let us have eternal thoughts circulating among us as well as gold and silver."[62]

Science and technology were advancing at a rapid pace, and Tappan wanted this country's intellectual resources to be nurtured and advanced at a comparable rate.

The railroad first reached Ann Arbor on October 15, 1839 when John Monroe completed construction of a track from Ypsilanti to Ann Arbor, with the first passenger car arriving in Ann Arbor from Detroit two days later. [Figure 3.2] It was a momentous day. Members of the Detroit City Council and about

Figure 3.2. Train of the same type as the first to reach Ann Arbor in 1839. (Silas Farmer, 1884)

800 citizens boarded a train and headed to Ann Arbor, where nearly the entire population gathered to greet them with a tremendous ovation, and a feast for all was laid out in the Court House Square.[63] The Michigan Central Railroad, which began as the Detroit St. Joseph Railroad, had been chartered in 1832. Colonel John Berrien, under direction of the War Department, had surveyed the route for its track. Initial progress was slow: in 1838 there had been only 38 miles of track in Michigan, and by 1845, the track extended only from Detroit to Marshall. By 1852, however, when Henry Tappan was inaugurated as President of the University there were 425 miles of track. Construction again stagnated for a few years, but in 1855 construction began again, and continued steadily until, by 1884, there were 5,120 miles of track.[64]

A telegraph line was strung along the right of way of the Michigan Central Railroad's tracks in 1847-48 by Jeptha P. Wade, an itinerant painter who found his calling as an employee of telegraph entrepreneur Ezra Cornell. Wade successfully obtained the reluctant consent of railroad engineer J. W. Brooks[65] to string the telegraph line. Brooks expressed skepticism regarding the ability of the telegraph to assist his railroad: "Why, I had rather have one hand car for keeping my road in repair and handling my trains than all the telegraph lines you can build."[66] The telegraph had to earn the confidence of men like Brooks. It was a few years before telegraph lines worked with sufficient reliability to gain the confidence of the railroad, but once the benefits were realized, the telegraph led to a revolution in American railway operation. Wade's success in his work with Cornell led him to become president of the Cleveland & Cincinnati Telegraph Company in 1854.

Henry N. Walker had important connections to the railroad that distinguished him in the early history of Michigan. He was a prominent lawyer who began his legal studies in New York State, moving to Michigan in 1834. He finished his legal training in the law office of the eminent attorney, Elon Farnsworth, who served several terms as a Regent of the University of Michigan between 1837 and 1857, and played an active role in recruiting Tappan. Walker developed the reputation as one of the best equity lawyers in the State, and he distinguished his career through his involvement in the early organization and construction of the railroad in Michigan.

It was in 1845 that Henry Walker, who was then attorney general of the State of Michigan, was summoned to negotiate the sale of the Michigan Central Railroad (MCRR). At a meeting held in the City Hotel at Albany, a new charter was drafted with Erastus Corning, who held a large share of the State of Michigan bonds issued for the MCRR, and with J. W. Brooks, who was superintending a rail line between Rochester and Syracuse, New York. The MCRR was to become the property of private interests. In March of 1846, with the assistance of George E. Hand, then a member of the Legislature, an Act was passed for the incorporation of the MCRR Company, and for the sale by the State of its interest, for the sum of $2,000,000.

Walker's success in the Michigan Central Railroad negotiation helped him gain a reputation as a skilled railroad lawyer, which led to his involvement in many future railroad deals. He became personally invested in the railroad business in 1848 when he became president of the Oakland & Ottawa Railroad, which was an extension of the Detroit & Pontiac Railroad across the State of Michigan to Lake Michigan. Walker became expert at securing subscriptions (fundraising) for his railroad endeavors, and in the process, he became the business associate of many leading railroad men, including Corning, Brooks, Farnsworth, and James F. Joy.[67] Walker was also a prominent Detroit bank official.

With his excellent fundraising credentials in hand, Walker pledged to work with Tappan to secure subscriptions to erect an observatory. The association between Walker and Tappan was to be a magical match of scientific and applied interests: Tappan would have his scientific laboratory for the advancement of knowledge, and Walker's railroad would benefit from the applied uses of astronomical science. For both of these educated men, the act of advancing science in the State would be highly gratifying.

As a first step in fundraising, Henry Walker called together on the evening of December 29, 1852 a group of his business acquaintances, prominent citizens of Detroit, scientific men, members of the Detroit Young Men's Society (a group devoted to academic and scientific advancements), and others at the Michigan Exchange, a hotel in Detroit,[68] to discuss fundraising for an observatory. Tappan unfolded the details of his idea for an observatory, and several others made speeches in support of the idea. The group was enthusiastic in their response, quickly raising an initial $7,500 in subscriptions, with the understanding that at least $10,000 would be raised. In the President's Report to the Regents in October 1854, Tappan recognized these donors as "friends of Science in the City of Detroit." About the observatory, Tappan said:

> This enterprise will prove of great advantage not only to the cause of Science but to the University, and the thanks not only to those immediately connected with the Institution, but of the State, are due to the intelligent and liberal donors.[69]

In honor of the beneficence of the many generous citizens from Detroit who stepped forward to contribute toward the project, the observatory was named the Detroit Observatory. The City of Detroit would reciprocate in 1867 by honoring Henry Tappan: Detroit's new Tappan School was erected at the corner of 13th and Marantette Streets. [Figure 3.3]

Figure 3.3. Tappan School. (Silas Farmer, 1884)

Tappan's original vision was to build a modest structure that would house a refracting telescope to be used to advance scientific knowledge through instruction, and to further scholarly research.[70] But, Walker set his sights on other coordinates. It was Walker who suggested the additional acquisition of a meridian circle telescope of the finest quality, an instrument essential for the first-class observatory Walker envisioned. So much did he desire this telescope, he financed its entire cost of $4,000 (approximately $60,000 in 1998 dollars).

A meridian circle telescope, and an astronomer skilled in its use, would provide Henry Walker and the citizens of Detroit with accurate time. This would be invaluable to railroad, mining and lumber interests, and to other business endeavors in which Walker and many of his Detroit associates were invested. It had just been announced in 1851 that electro-magnetic induction was capable of making instantaneous registration of astronomical observations,[71] and this would have been exciting news to the Detroit donors, with direct application to running a railroad, surveying land, and other business endeavors.

The railroads served many business interests. Mining speculators teamed with railroad entrepreneurs in order to transport minerals from Upper Peninsula mines to market locations across the State and beyond. The same held true for trappers who needed to transport furs and skins to market locations for the manufacture of shoes, hats, and other leather goods. Similarly, lumbermen needed to transport the trees they harvested so that mills throughout the region could provide lumber, which was in great demand.

Copper mining fever was rampant in Michigan in the 1850s. Native Americans had mined copper for many centuries. Douglass Houghton had been confident, based upon first-hand explorations in the 1830s, that copper was abundant in the Upper Peninsula of Michigan. As geologist for the State, Houghton issued a report in 1841 that captured wide attention. As a result, hundreds of geologists and explorers traveled to the Lake Superior region from all over the world. Pressure was placed on the Government to get the area surveyed, and Houghton was assigned to begin the work in 1844. Copper fever spread quickly. Even the Tappans were swept up in copper investments. Mrs. Tappan's aunt, Mrs. Noon, who lived with the Tappans in Ann Arbor, purchased some Lake Superior copper land through Regent Charles H. Palmer soon after her arrival in Ann Arbor.[72] After leaving Michigan for Europe, Tappan wrote to Palmer, enquiring about the copper land belonging to Mrs. Tappan, who had apparently inherited the claim when Mrs. Noon died in 1861.[73] Mrs. Tappan refused to sell the land, and Tappan regularly enquired by letter about any property taxes due, and sent payments to ensure continued ownership.[74] From 1850 to 1877, Michigan copper mines were the most important and productive in the nation. Iron ore was discovered at Negaunee in 1844, which added further dimension to the copper mining boom already in progress. By 1846, iron ore was being shipped from Marquette. The Upper Peninsula had very quickly become a busy industrial center.

Beyond their financial interests in establishing an observatory, many of the donors to the Observatory had an intellectual interest, as well. Henry Walker and some of the other donors were members of the Detroit Young Men's Society,[75] a group formed as part of the lyceum (cultural) movement that was long considered to be "the most powerful intellectual force in Michigan."[76] The lyceum movement was strong in New England, and it migrated to Michigan as New Englanders took increasing advantage of the Erie Canal and steamboat passage to Detroit, and later by railroad.

Several of the donors to the Observatory served as president of the Detroit Young Men's Society. For members of the Society, beyond advancing their business interests, their contributions toward the establishment of an observatory would be consistent with their desire to promote intellectual and scientific advancements in the State. Douglass Houghton, the natural scientist, surgeon, and geological surveyor, had helped to establish the organization in 1832. The Society served as the City's center of higher learning, with its extensive library holdings, eventually numbering sixteen thousand volumes at its peak, serving as the group's intellectual fuel. (When it became apparent in 1882 that the Detroit Public Library had surpassed the Society's holdings, the Society disbanded.) Throughout its active life, the Society hosted a lecture series that featured distinguished authors and orators, including Ralph Waldo Emerson, Edward Everett, Horace Greeley, and Mark Twain. The Society was a training-ground for governors, senators and judges.[77] One success story was Henry P. Baldwin, a donor to the Observatory who was a banker, business owner, senator, and eventually governor of the State.

A close examination of the characteristics of the initial group of Observatory donors is helpful in understanding their enthusiasm. Altogether, there were twenty-nine in this group. Leading the list as the most generous, with pledges of $500 (approximately $7,500 in 1998 dollars) were Henry Walker; Lewis Cass, [Figure 3.4] American statesman, Governor of Michigan from 1813-31, U.S. Senator, candidate for President of the United States in 1848, Indian agent, and explorer with Henry Rowe

Figure 3.4. Lewis Cass portrait by Alvah Bradish, pictured with telescope, books, and globe as symbols of his prominent status. (Michigan Historical Museum)

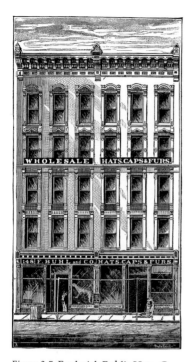

Figure 3.5. Frederick Buhl's Hats, Caps and Furs store, built in 1852, Detroit. The Buhls were donors to the Detroit Observatory. (Silas Farmer, 1884)

Schoolcraft to survey the upper lakes of the Mississippi; Henry P. Baldwin, prosperous shoe merchant, was to become one of the directors of the Second National Bank of Detroit, President of the Detroit Young Men's Society, a member of the U.S. Senate, and eventually the Governor of the State; Zachariah Chandler, a wealthy dry goods dealer who was Mayor of Detroit and the Whig nominee for Governor of Michigan, later became a U.S. Senator, and had his business home at the Second National Bank; Frederick and Christian H. Buhl,[78] who owned a successful hat business [Figure 3.5] (later opening an iron works and a hardware business), had their business home at the Second National Bank, and both served as mayors of Detroit; J. W. Brooks, an eminent railroad engineer and Vice President of the Sault Canal Company that built a ship canal around St. Mary's Falls, thereby permitting freighters to transport ore from the Upper Peninsula; E. C. Litchfield,[79] a wealthy lawyer and banker who spent his career in railroad enterprises and real estate development, in Detroit as President of the Farmer's and Mechanics Bank from 1845-52, and then relocating to Brooklyn, New York, was related to another donor, Bela Hubbard, through his marriage to Hubbard's sister, Grace Hill Hubbard Litchfield; Buckminster Wight, a land investor for whom another donor, H. David A. Ward, surveyed and selected investment property; Edmund A. Brush, who was related to General Cass through marriage to Cass' daughter, had accompanied Cass on his exploration for minerals in 1826, and was later appointed by Cass as the City Register for Detroit, taking a considerable interest in the development of the railroad;[80] Smith, Dwight & Co, and Catharine H. Jones.

Other influential citizens joined in, with contributions of $100 (approximately $1,500 in 1998 dollars), except where noted: Bela Hubbard, a highly-successful land investor, scientist, explorer, writer,[81] lumberer, lawyer, and philanthropist who served as assistant geologist on Douglass Houghton's geological survey of the State, being ever watchful for potential land deals and mining investments; Samuel T. Douglass, a judge and brother of Silas H. Douglass who was a professor of chemistry at the University, accompanied Douglass Houghton and Bela Hubbard as doctor on the geological survey of Michigan, and superintended the construction of the Detroit Observatory;

D. Bethune Duffield and Henry Ledyard of the Ledyard & Duffield law firm that represented railroad interests, Duffield being the son of George Duffield, a Regent of the University from 1839-48 and pastor of the Presbyterian church,[82] and Ledyard (a descendant of Robert Livingston) who first met Lewis Cass on a ship bound for Europe and eventually moved to Detroit as Cass' son-in-law and as a special commissioner involved with the ship canal around St. Mary's Falls; Shubael Conant, a merchant, furrier, and the builder of the Michigan Exchange in Detroit, where the donors held their initial meeting; John Owen, a former Regent of the University, banker, and President of the Detroit and Cleveland line of steamers; Charles C. Trowbridge, politician, President of the Michigan State Bank and leader of the Milwaukee Railroad Company; and H. David A. Ward, a surveyor who started the custom of receiving one-quarter of the land for exploring and making land selections for patrons. Other donors included C. A. Trowbridge ($100); W. M. Whitcomb ($100); J. W. Tillman ($100); E. N. Wilcox ($50); Franklin Moore ($250); S. Barstow ($100); J. A. Van Dyke ($200); S. N. Kendrick ($100); C. Howard ($100); and Duncan Stewart.

With this first round of pledges in hand, President Tappan moved forward, with the understanding that a minimum of $10,000 would be raised. The success of this first effort gave Tappan and Walker confidence that they could proceed with their expanded plan for a first-class observatory. Tappan headed off to purchase the telescopes for the building, first to New York and from there to Europe. A second fundraising effort, in May of 1854, was successful in garnering another $4,300, which included an additional $3,200 from Henry Walker. Other donors at the $100 level included Moses W. Field, owner of a large mercantile house; Detroit Mayor John H. Harmon; W. S. Driggs; J. W. Brooks; and merchant and banker Theo. H. Eaton. Pledges of $50 were made by E. W. Hudson; T. H. Hinchman; John Winder, a Clerk of the U.S. Circuit Court; Eliza Stuart; Pittman, Trowbridge & Jones; C. A. Trowbridge; and H. Doty. In addition, $25 pledges were made by J. H. Hickson; J. H. Armstrong; S. C. Whittmore; S. Larned; E. A. Lansing; two members of the Noyes family; J. C. Holmes; lawyer James V. Campbell; and Bridge & Lewis.

By 1856, the Observatory was completed and operational, but a balance remained outstanding. Tappan was being pressured by University officials to close the account on the Observatory, so much so that an audit of the books was demanded. Henry Walker called another meeting of the subscribers to the Detroit Observatory. They met at the National Hotel [Figure 3.6] on March 13th where

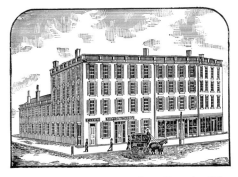

Figure 3.6. National Hotel in Detroit. (Silas Farmer, 1884)

they expressed their entire satisfaction with the manner in which President Tappan had executed their trust through the successful completion of the Observatory.

In order to expand the subscriptions pledged to the Detroit Observatory to liquidate the balance still due, a committee was formed, comprised of Tappan, Frederick Buhl, Theodore Eaton, C. I. Walker, and C. A. Trowbridge. This group was successful in gaining $3,510 in subscriptions, which, along with some funds already pledged by the University, satisfied the balance due and freed Tappan from the Regents' scrutiny. The total cost of the project was about $22,000. Lewis Cass gave another $500 contribution, and Frederick Buhl provided an additional $125; Theodore H. Eaton gave another $100; James F. Joy, a prominent lawyer and former professor whose successful handling of significant railroad cases led to his direct involvement in the railroad enterprise through association with John W. Brooks (another of the donors), gave $500; Alanson Sheley, a wholesale druggist who was also in the lumber business, using the railroad to transport his goods, contributed $500; Elon Farnsworth, a University Regent, lawyer, Chancellor of the State for the brief time that office existed, banker, Director of the Michigan Central Railroad, and legal agent for the international sale of land owned by the Sault Canal Company, gave $200; Silas M. Holmes ($200), owner of a major dry goods store; George E. Hand ($100), a judge and personal friend of Lewis Cass, had oversight of the bills for chartering the Michigan Central and Michigan Southern Railroad companies; Philo Parsons, a capitalist with roots in the wholesale grocery trade, and a staunch supporter of higher education, gave $100; and George B. Russel, who organized a company in Detroit to manufacture railroad cars (later acquired by George M. Pullman), donated $100. Other donors included J. W. Waterman ($500); D. Cooper ($200); W. A. Butler ($100); Walker & Russell ($100); Campbell & Linn ($60); W. N. Campbell ($25); and Detroit merchant A. C. McGraw ($100).[83]

Tappan's social, educational, and political connections, and his personal warmth and charm, undoubtedly contributed to the success of his efforts at fundraising for the Observatory. Prominent families of University students were included in Tappan's fundraising bailiwick. In 1856, Alexander C. McGraw, a shoe merchant in Detroit, wrote to his son, Edward, who was a student at the University, that President Tappan had been to see him in Detroit. [Figure 3.7] McGraw reported that he had donated $100 toward the Observatory, and gave Tappan a pair of boots for himself. McGraw found Tappan to be "remarkably social and friendly," and he used Tappan as an example to his son of the personal characteristics he ought to emulate in order to advance himself.[84] Mrs. Tappan also got involved in fundraising activities. In 1858, she undertook a fundraising campaign, collecting a little over $200 in $1 pledges from friends and acquaintances, mostly in Ann Arbor, toward the support of the Mt. Vernon Fund.[85]

Figure 3.7. Detroit residence of Alexander C. McGraw, father of a University of Michigan student, where Tappan paid a visit in 1856 to raise funds for the Detroit Observatory. (Silas Farmer, 1884)

Like the Detroit donors, the citizens of New York State were motivated to create an observatory in the early 1850s, but the singular basis for their motivation was the practical benefit an observatory could provide to the City of Albany, and to New York State. The observatory they envisioned would not be connected to a university but would, instead, be limited to providing the applied needs of the State. In contrast to Tappan's plan for a refracting telescope that would serve instructional and research purposes, the Trustees of the Dudley Observatory in Albany wanted only a meridian circle telescope. It was O. M. Mitchel, the astronomer from the Cincinnati Observatory who was engaged, in absentia, to direct the proposed Dudley Observatory, who very gently encouraged the Dudley Trustees to consider a design for the new observatory that could eventually accommodate a refracting telescope in order to meet "popular demand." Yet, Mitchel further suggested that the refracting telescope should be an instrument of equal size to the 12-inch refractor of the Cincinnati Observatory.[86] Mitchel's interest in academic astronomy, not simply the applied uses of astronomy, is quite evident. Mitchel's persuasion was eventually successful, and a 13-inch refracting telescope by Henry Fitz was installed at the Dudley Observatory in 1860.

It wasn't until December 1858, four years after Franz Brünnow arrived in Ann Arbor from Berlin to become the first director of the Detroit Observatory, that the citizens of Detroit asked him to investigate how the correct time could be telegraphed from the Observatory to the City of Detroit.[87] Time was kept for Detroit by jeweler Martin S. Smith. It was common for time to be kept by jewelers in the early days, given their familiarity with timepieces. In 1841, William A. Burt provided Smith with the newly-invented Burt's Sun Dial, which was mounted in front of Smith's store and used to regulate Smith's public clock.[88] Later, Smith had a small transit telescope mounted on his rooftop.[89] In Brünnow's estimation ". . . I do not think he knows much about observing, nor does it seem he makes observations very often. There are constant complaints in Detroit about the time . . ."[90]

In 1863, a Detroit newspaper reported that "by telegraph, mean time is furnished to neighboring cities, especially Detroit and Milwaukee. Our time is 2 min. 43 sec. faster than that of Ann Arbor and our mean time, as determined by the Observatory, is kept by Smith on Woodward Avenue."[91] Outside Smith's jewelry store was displayed a large clock surmounted by a bronze eagle, used by passersby to set their watches. Atop Smith's five-storey building was mounted a timeball, which could be seen from a distance, by which clocks and watches throughout the City could be set. [Figure 3.8] The ball was raised just before noon, and dropped at precisely noon[92] according to a signal sent by telegraph from the Detroit Observatory.

The use of a timeball originated in 1833 at the Greenwich Observatory near London, England.[93] Today, the large, red timeball at Greenwich is still dropped daily, precisely at 1:00 p.m.—quite a spectacle to behold. [Figure 3.9] This time signal was used by ship captains in Greenwich harbor to set their chronometers before sailing out to sea. Time was a critically important element of sailing because accurate time was essential to navigation.

In response to the 1858 request from the citizens of Detroit, Brünnow enlisted the assistance of William C. Bond of Cambridge Observatory at Harvard College to provide the details needed to establish a time service in Michigan, based on Bond's experience in providing the time for the City of Boston. A telegraph line was strung from the Detroit Observatory, directly west to the property line, then turning abruptly

Figure 3.8. Timeball atop Smith's jewelry store, and corner clock, were used to keep the time for Detroit. (Silas Farmer, 1884)

Figure 3.9. Timeball at Greenwich Observatory near London was used to signal the time to ships in Greenwich harbor so ship chronometers could be set to the correct time before departure to sea. The timeball is still dropped daily at 1:00 p.m., shown here at the mid-point of its descent. (Patricia S. Whitesell)

north to the railroad tracks, and along the tracks to the west until the line reached the Michigan Central Railroad Depot.[94] [Figure 3.10]

The exact date at which the Detroit Observatory commenced sending time signals to Detroit by telegraph is uncertain. On October 1, 1860, Brünnow wrote to G. P. Bond at Harvard that a telegraph connection was established between the Observatory and the Lake Survey at Detroit, but this does not refer to the establishment of a time service.[95] What the Lake Survey attempted to do was determine longitude for their surveying work by comparing their time by telegraphic means with that of the Detroit Observatory. Unfortunately, this work was attempted during the time Brünnow was absent from Ann Arbor for a little over a year as the Associate Director of the Dudley Observatory in Albany (see Chapter 12). Watson, Brünnow's student assistant, who was left in charge of the telescopes, failed to make the required observations, so the Lake Survey had to wait two or three weeks before Watson provided the reductions needed to verify the time.[96]

When Brünnow returned to his position in Ann Arbor in 1860, he obtained approval from the Board of Regents to purchase a chronograph, made by the Bonds of Harvard College, for $360 (approximately $4,500 in 1998 dollars).[97] [Figure 3.11] The chronograph was shipped to Ann Arbor, but was delayed in

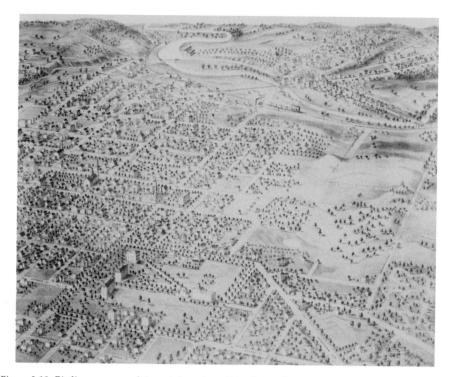

Figure 3.10. Bird's eye map of Ann Arbor circa 1866 shows the telegraph line that ran from the Detroit Observatory (right edge, center) to the train station. (Bentley Historical Library)

transit. Brünnow wrote to the Bonds, "the chronograph has not arrived yet probably on account of the crowded state of the Canal."[98] It finally arrived in September. The chronograph was an instrument used to record the position of stars as they crossed the meridian—data essential to determining accurate time. The chronograph consisted of a brass drum covered with paper, and a pen connected to a galvanic battery. The cylinder revolved once a minute, and it was connected by electrical wires to a sidereal clock. Each swing of the pendulum broke the electric current, thereby jogging the pen to record the seconds on the paper. The observer, looking through the meridian circle telescope, would record the position of stars by pushing a control that made the pen record the star's position on the paper.[99] Brünnow and other astronomers at the Detroit Observatory kept detailed log books into which they entered data collected in their observations with the meridian circle telescope. [Figure 3.12]

Brünnow had learned a great deal about the process of timekeeping while he was at the Dudley Observatory in Albany, where he had participated in putting a time service into operation. Yet, Brünnow's efforts to provide a time service in Ann Arbor were surely driven by the Detroit benefactors' request rather than by Brünnow's independent interest in the pursuit. His personal interest in astronomical science was strictly academic.

On July 16, 1857, the first telegraphic cable was laid across the Detroit River. To achieve the durability needed for such a feat, a cable identical to the Atlantic cable used at Newfoundland[100] was obtained. It was considered to be "the first really successful submarine telegraph cable laid in any waters."[101] A puzzling entry in the University's *Regents Proceedings* may be connected with this event: in September 1858, the Regents approved payment of $12.63 (a little less than $200 in 1998 dollars) to Tiffany & Co. for twenty feet of Atlantic cable. Exactly how the University used this cable is not known.

With the telegraph cable in place across the Detroit River, and the telegraphic connection made with the Lake Survey (1860), Captain George Meade of the Lake Survey proposed to Brünnow that he establish the exact longitude of the Detroit Observatory, and Brünnow eagerly accepted. Brünnow arranged a collaboration in 1861 with fellow German astronomer C. H. F. Peters at the Hamilton College Observatory in Clinton, New York (near Utica) to determine the longitude. The longitude of Ann Arbor had previously been determined by Brünnow using occulations of the stars of Pleiades, but a much more precise determination could now be made by comparing the time of the two observatories using the chronographs and the telegraph line. This endeavor was complicated by the Civil War because the telegraph lines were in use throughout the night, but the superintendents of the New York Central Railroad and the Western Telegraph allowed the astronomers to use one wire for two nights, after midnight.[102]

The first connection between the Detroit Observatory and Hamilton College Observatory was made on June 29, 1860 when the mean time clock at Hamilton College and the sidereal clock at Ann Arbor were transmitted over the telegraph line, with 126 beats of the two clocks recorded on the two chronographs. [Figure 3.13] It must have been thrilling for the astronomers to hear the beating of a clock such a great distance away. A second successful connection was established on July 3rd. The data gathered were corrected for errors in the precision of the clocks, in collimation and azimuth using the method of least squares, and in level as determined by the spirit level on the meridian circle. The result was a

Figure 3.11. Chronograph by Fauth & Company is still extant at the Detroit Observatory. (Patricia S. Whitesell)

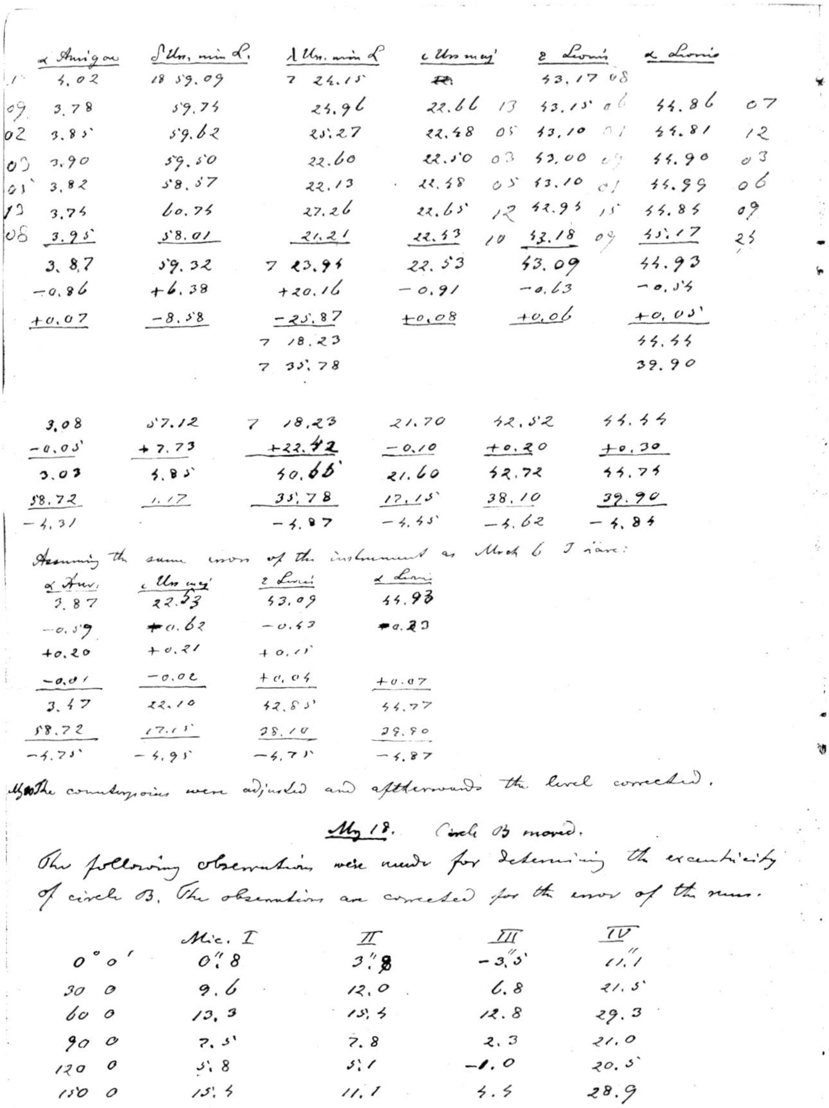

α Aurigae	δ Urs. min ♃	λ Urs. min ♃	ε Urs. maj'	ε Leonis	α Leonis
4.02	18 59.09	7 24.15		43.17 08	
3.78	59.74	24.96	22.66 13	43.15 06	44.86 07
3.85	59.62	25.27	22.48 05	43.10	44.81 12
3.90	59.50	22.60	22.50 03	43.00	44.90 03
3.82	58.57	22.13	22.58 05	43.10	44.99 06
3.74	60.75	27.26	22.65 12	42.95 15	44.85 09
3.95	58.01	24.21	22.53 10	43.18 09	45.17 25
3.87	59.32	7 23.94	22.53	43.09	44.93
−0.86	+6.38	+20.16	−0.91	−0.63	−0.54
+0.07	−8.58	−25.87	+0.08	+0.06	+0.05
		7 18.23			44.44
		7 33.78			39.90

3.08	57.12	7 18.23	21.70	42.52	44.44
−0.05	+7.73	+22.42	−0.10	+0.20	+0.30
3.03	4.85	40.65	21.60	42.72	44.74
58.72	1.17	33.78	17.15	38.10	39.90
−4.31		−4.87	−4.45	−4.62	−4.84

Assuming the same error of the instrument as March 6 I have:

α Aur.	ε Urs. maj'	ε Leonis	α Leonis
3.87	22.53	43.09	44.93
−0.59	+0.62	−0.43	−0.23
+0.20	+0.21	+0.11	
−0.01	−0.02	+0.04	+0.07
3.47	22.10	42.85	44.77
58.72	17.15	38.10	39.80
−4.75	−4.95	−4.75	−5.87

May the comparisons were adjusted and afterwards the level corrected.

May 18. Circle B moved.

The following observations were made for determining the eccentricity of circle B. The observations are corrected for the error of the runs.

	Mic. I	II	III	IV
0° 0'	0".8	3".8	−3".5	11".1
30 0	9.6	12.0	6.8	21.5
60 0	13.3	15.5	12.8	29.3
90 0	7.5	7.8	2.3	21.0
120 0	5.8	5.1	−1.0	20.5
150 0	15.5	11.1	4.5	28.9
180 0	6.5	2.1	53.2	15.2
210 0	11.2	7.9	57.6	18.0
250 0	11.7	11.5	56.2	16.5
270 0	−5.1	−3.0	−16.4	2.0
300 0	6.8	11.2	−1.9	13.2
330 0	5.3	9.9	−0.5	13.1

Figure 3.12. Franz Brünnow recorded in his log book on May 18, 1855 the observations he made to determine the eccentricity of circle B of the Pistor & Martins meridian circle telescope. (Bentley Historical Library)

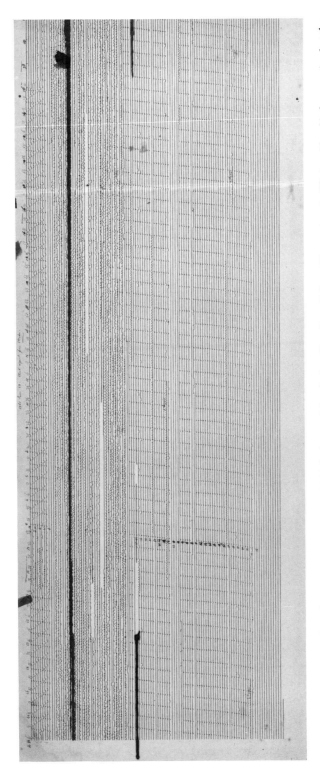

Figure 3.13. Chronograph sheet that recorded on June 29, 1861 the beats of the clock at Hamilton College Observatory in Clinton, New York as received at the Detroit Observatory by telegraphic connection. These data were used to determine the exact longitude of Ann Arbor. (Detroit Observatory Collection)

successful determination of the longitude of the Ann Arbor meridian circle, set at $50^m24^s.21\pm0.047$ west of Cambridge or $5^h34^m54^s.87$ west of Greenwich. This was later corrected to $5^h34^m55^s.27$ west of Greenwich after the value for the Harvard observatory's longitude was corrected. Brünnow also determined the latitude of the Detroit Observatory, the value of which was extremely close to the present-day accepted value.[103]

The determination of the longitude and latitude of the Detroit Observatory was extremely valuable: all subsequent surveys would be corrected to this new level of accuracy; all future surveys would benefit from an accurate starting reference point; and all future longitude determinations to the west would use the longitude of the Detroit Observatory as a reference point.

In 1864, the Regents were requested to authorize funds to repair the telegraph connected with the Observatory,[104] which confirms that telegraph lines were operational before Brünnow resigned in 1863 coincident with President Tappan's dismissal from the University [see Chapter 13]. Many observatories used income from providing the time to support operations. No records have been located to indicate there was income to the Detroit Observatory for providing the time, but it was not uncommon for astronomers to keep such revenue themselves. Given James Craig Watson's obsession with the pursuit of personal gain, [see Chapter 11] it is possible that Watson, the second director of the Detroit Observatory, continued the time service Brünnow established, and personally kept any revenues.

Over its history, the Detroit Observatory served numerous scientific and applied needs, not limited to astronomy. Tappan believed that all of the sciences were historically connected with astronomy, and this was to be true of the life of the Detroit Observatory. Numerous disciplines were represented by the various directors and faculty appointed at the Observatory over the years, through instructional and research activities, and through their personal interests. To mention a few examples, Brünnow brought mathematics, rigorous academic standards, international culture, native fluency in German, writing skills, and a publication ethic, as well as his considerable musical talents; Watson was engrossed in business and financial endeavors, and he published detailed tables he calculated of compound interest; Mark W. Harrington, the Observatory's third director, was expert in botany and natural history, and he intensified the Observatory's meteorological activities, eventually leaving the University in 1891 to become the inaugural chief of the U.S. Weather Bureau; Asaph Hall, Jr., the fourth director, made the Observatory available to ornithologists who studied the relation of the weather to bird migration by observing birds as they passed between the telescope and the moon.[105] William J. Hussey, the fifth director, expanded efforts Harrington began in seismology by adding more sophisticated instruments to record and study earth tremors, and he established the Observatory Shop in 1905 so that University instrument makers could maintain the increasingly more sophisticated apparatus used for astrophysical work. In addition, librarians were trained using the Observatory's extensive collection of books and journals. The breadth and diversity of scholarly and applied activities undertaken at the Observatory over the years leaves behind a history that significantly enriches and adds breadth to historical study today.

Figure 4.1. Warehouse in New York Harbor where the meridian circle telescope was received from Berlin. Reed & Sturges became Sturges, Bennett & Co. circa 1853. Photograph by Victor Prevost. (Collection of the New York Historical Society)

Chapter 4
Pistor & Martins Meridian Circle

Tappan had planned, even before his inauguration as President, to travel abroad in 1853, and he followed through on his plan with the understanding that, while traveling, he would examine the European universities' system of higher education. With Henry Walker's funds in hand, he added to his itinerary the purchase of a transit telescope and an astronomical clock. In February of 1853, Tappan traveled to New York City with Walker, where they met at the home of Julia Tappan's father, Colonel John W. Livingston. Present were a number of scientific men and Henry N. Fitz, Jr., [see Chapter 5] with whom they would establish a contract for the construction of an achromatic refracting telescope with an equatorial mount at a cost of $6,150 (approximately $92,000 in 1998 dollars). Tappan specified that this telescope was to be least 12 inches clear aperture, which would make it second only to the great refractors at Cambridge (Harvard University) and Pulkova in Russia.

Having completed these arrangements, Walker returned to Detroit while Tappan made plans to sail for Europe. Before his departure, Tappan made an arrangement with Sturges, Bennett & Co. to have the meridian circle telescope shipped to their warehouse at 125 Front Street in New York Harbor. [Figure 4.1] They were then to send it on via canal and railroad to Ann Arbor, for which they were paid $149.25.[106] Mr. Sturges was a prominent merchant, co-founder of the Illinois Central Railroad, and patron of the Hudson River School of artists. The Sturges family desired to send their two children, Virginia and Fred, to Europe, and upon learning that Tappan was planning a trip there to purchase instruments for the Observatory, they "proposed putting the children under his care, for a consideration, in order that they might make the journey under proper auspices." Tappan accepted the proposal, and decided that Mrs. Tappan and their daughter, Rebecca, would accompany him: he had cleverly financed their passage. They sailed on the American steamer Atlantic with Captain West, "one of the best of the old Collins line of steamers, entirely under American auspices." Their departure was on a Saturday night in February in the midst of a windy, winter storm, but all was clear by morning.[107]

They arrived safely in England where Tappan toured the Greenwich Observatory. There he consulted with Professor Airy regarding that observatory's new transit instrument but, finding it to be too large and expensive in its construction, Tappan abandoned any idea of using it as a model. In Rome, Tappan toured the Observatory of the Roman College, which was under the direction of Professor Secchi, who was in the midst of making alterations to the Observatory in preparation for new instruments that had been ordered from Oertel, an instrument maker in Munich. Tappan then traveled to Switzerland through the Simplon Pass and left Mrs. Tappan and Rebecca in Geneva.

Tappan traveled on by himself to Berlin before going to Munich. There he met Professor Encke, a most distinguished astronomer of the Royal Observatory in Berlin, whose views on transit instruments were in agreement with Secchi's. When asked, Encke advised that his instruments were always made by the firm of Pistor & Martins of Berlin, and he recommended them unequivocally "and yet with characteristic delicacy lest he should interfere with any previous arrangement" Tappan may already have made. Tappan indicated he would look no further, placing an order with Pistor & Martins at once, if Encke would supervise the construction of the instrument.[108] Encke and his assistant, Dr. Franz Brünnow, agreed and the following contract was drawn up:

> We do hereby engage to make for the University of Michigan, in the United States of America, a Transit instrument [Meridian Circle] with a telescope of eight feet focal length, English measure, with an object glass of seventy-two French lines in diameter, with two divided circles of three feet diameter each, with eight microscopes, and complete furniture throughout as described under number one of our *Preis Verzeichniss*. We engage to furnish and deliver the same by May 1st, 1854, and to pack and forward the same to New York to the care of *Messrs.* Sturges, Bennett & Co., unless in the meantime otherwise directed by the Chancellor of the above named University. We accept Professor Encke and Dr. Brünnow of Berlin, as the judges of the instrument and engage to furnish one with which they will be satisfied. The above instrument we engage to make for the sum of four thousand Prussian thalers to be paid to us upon delivery of the instrument.

> Dated Berlin, July 15th, 1853

> In witness whereof we have set the hand and seal of our firm
>   [Signed]
>   Pistor & Martins.

Witness [Signed] Dr. Brünnow[109]

Because the instrument was not to be paid for until received in Ann Arbor, Tappan returned Henry Walker's draft to him, whereupon Walker paid for the instrument on arrival and donated it to the University. A brass plaque was placed on the meridian circle's stone pier to identify the telescope as the gift of Henry N. Walker of Detroit, and the instrument was referred to as the Walker meridian circle. [Figure 4.2]

Figure 4.2. Plaque on meridian circle telescope at the Detroit Observatory honoring Henry N. Walker of Detroit who financed the instrument's purchase. (Patricia S. Whitesell)

Figure 4.3. Sidereal clock (still extant) made in 1854 by M. Tiede of Berlin. (Patricia S. Whitesell)

Figure 4.4. Stereoscopic photograph of the meridian circle telescope at the Detroit Observatory, made in 1854 by Pistor & Martins of Berlin. The face of the sidereal clock by Tiede can be seen mounted on the clock pier behind the left circle, with the clock's case extending down below the eyepiece. When viewed through a stereoscope, the image appears in three dimensions. (Bentley Historical Library)

Before leaving Berlin, upon Encke's recommendation, Tappan ordered from M. Tiede at a cost of $300 an astronomical clock rated to sidereal time.[110] Tiede had made a beautiful astronomical clock for the Berlin observatory, with which Drs. Encke and Brünnow had been exceptionally pleased; they, no doubt, recommended Tiede's horological artistry for Dr. Tappan's observatory. [Figures 4.3 and 4.4] After returning to Ann Arbor in November, Tappan placed another order with Pistor & Martins for two collimating telescopes for $275. These collimators were needed to align the highly precise meridian circle instrument.

Carl Philipp Heinrich Pistor, who was born in 1778, began his career in 1793 with the Prussian Post Office. Pursuing his personal interests, he made astronomical observations with which he determined exact latitude and longitude. He went on to apprentice with mechanic K. T. N. Mendelssohn, and in 1813, he founded his own instrument making company in Berlin. In 1833, he took on as an apprentice Carl Otto Albrecht Martins, born in 1816, who later married Pistor's daughter. As partners, they manufactured meridian circle telescopes for the major German observatories (Leipzig, Bonn, and Berlin), and also for Leiden, Copenhagen, Palermo, the U.S. Naval Observatory in Washington, the Amherst College Observatory in Massachusetts, and the Dudley Observatory in Albany.[111] In addition, they made a meridian circle for the U.S. Naval expedition to Chile in 1849, and for the Dunsink Observatory outside Dublin at the request of Franz Brünnow, who was director there from 1865-74. The Pistor & Martins workshop closed in 1876.

Of the three Pistor & Martins transit instruments Brünnow would personally install [Detroit, Dudley, and Dunsink observatories], only the Detroit Observatory's instrument remains intact in its original mount. In fact, Michigan's Pistor & Martins instrument is the oldest surviving meridian circle in America, and the oldest surviving Pistor & Martins meridian circle still in its original mount in the entire world. A nearly identical telescope they made in 1856 for the Dudley Observatory is currently in storage at the New York State Museum's warehouse in Rotterdam, New York, but only pieces survive: the two circles, lamps, mercury basin, levels, brake arm, and the empty brass tube, which is badly damaged. The 1865 meridian circle telescope of the U.S. Naval Observatory is no longer extant, nor is the Dunsink instrument, except for its two circles; only the reticle remains from the Amherst College meridian circle. The instrument made in 1860 for Leiden is on display at the Boerhaave Museum for the History of the Natural Sciences in Leiden, The Netherlands, and the circa 1845 meridian circle from Bonn, Germany is currently on display at the Deutches Museum, Munich. The meridian circle made in 1859 for Copenhagen is currently on display at the Steno Museum, Århus, Denmark, and the instrument made for Palermo Observatory in 1854 has been modified nearly beyond recognition. The meridian circle made for the Berlin Observatory in 1838 was destroyed except for its objective, which is at the Astrophysical Institute, Potsdam-Babelsberg along with a second instrument made in 1867 for Berlin which is currently in poor condition.

Figure 4.5. Mercury basin, narrow gauge track for the observer's couch, track under floor panels, and stairs that enabled the astronomer to read the meridian telescope's circles by looking through the microscopes. (Patricia S. Whitesell)

The meridian circle telescope that Pistor & Martins made for the Detroit Observatory in 1854 had four positive eye pieces, with magnifying power ranging from 85 to 288 times. [Figure 4.4] The horizontal axis was perforated so that light emanating from oil lamps on either side and passing through the center of the telescope tube would be reflected to the eye by a mirror placed at a 45-degree angle, thereby illuminating the field of view. The mirror also contained perforations at its center to eliminate interference with the rays of light coming from the objective lens. The reticle contained "spider lines" set in a grid pattern, made of actual spider web filaments because of their fine texture and amazing tensile strength. These lines were illuminated and used by the observer to determine the exact timing of the passage of stars across the meridian. The circles were three feet two inches in diameter, mounted at the east and west sides of the telescope tube, with each circle containing an inlaid band of silver marked to two-minute intervals, and read by means of four microscopes to the nearest tenth of a second. The entire instrument was made of brass with the exception of the axis, which was bell metal, and the microscope arms, which were hardwood. Wooden stairs were constructed to permit the astronomer to read the circles.

The meridian circle was a highly precise instrument, with numerous devices provided to ensure its perfect alignment. Two collimating telescopes were used to align the telescope along the north-south meridian, and to detect any error in collimation, or flexure in the telescope's tube. Stairs were added around the stone piers to permit the astronomer easier access to the collimating telescopes. Because the collimating telescopes were purchased after the meridian circle arrived in Ann Arbor, these steps are of a slightly different style than those that serve the meridian circle. A spirit level, made of metal covered with thin leather, could be hung on the telescope, attached near each axis with the level's bubble in position directly below the tube. This was used to determine whether the instrument was properly aligned. When not in use, the level was stored on a wooden rack mounted low on the south wall to the west of the collimating telescope pier.

Figure 4.6. Reversing carriage for the meridian circle telescope rolls on a track on the floor. It is used to lift the telescope up off its mount and rotate it 180 degrees. (Patricia S. Whitesell)

Directly underneath the meridian circle telescope was a metal basin with a removable cover which, when filled with mercury, established a perfectly level mirror. [Figure 4.5] Reflections in this basin, called an artificial horizon, were used to determine the nadir and zenith, i.e., straight up and down. Over the course of nearly 150 years of use, the mercury poured into the basin wandered, as mercury has a tendency to do when spilled. Tiny beads of the silvery substance adhered to the metal track, and used the track to migrate the length of the room. Today, mercury is recognized as a toxic substance, so a University team of trained remediation technicians quickly and properly removed the hazard.

On a narrow gauge, railroad-style track that extended from north to south down the middle of the room ran an apparatus called a reversing carriage. [Figure 4.6] This device enabled the observer to hoist the telescope off its mount and rotate it 180 degrees, so that it was facing in the opposite direction from its previous orientation. This reversal made it possible to determine and correct for flexure in the instrument, thereby reducing the effect of instrument errors.

A couch for the observer also rolled on the track, enabling the observer to be in an optimal position relative to the eyepiece. [Figure 4.7] The couch was adjustable in its degree of incline by means of a simple wooden rod that could be moved forward or back between cogs, and it was reversible to permit observations to either the north or south. Another track of a narrower gauge ran between the larger track, but was hidden from view under hinged floor panels. On this track rolled a reflex table which could be raised and lowered like the reversing carriage. It held another mercury basin used as an artificial horizon for reflex observations. The track was laid on two solid wooden beams that extended between the telescope's piers but were totally insulated from

Figure 4.7. Observer's couch for the meridian circle telescope at the Detroit Observatory rolls on a track on the floor, and the angle of the seat back can be adjusted. Sun shining through the telescope lens caused the burn hole in the seat cushion. (Patricia S. Whitesell)

the floor, in order to eliminate any vibration that might create waves in the mercury.[112] A footfall in the room would otherwise have sent the mercury into convulsions, making it impossible to see a reflection.

The meridian circle telescope was very sensitive to changes in temperature, so the east wing in which it was located was purposely unheated, as was the dome room. A standard barometer and indoor and outdoor thermometers were included as part of the original equipment in the east wing.[113] These were used to ensure that the temperature to which the meridian circle was exposed was equalized, because changes in temperature could cause inaccuracies. A parasol was part of the original equipment, used to block sunlight that might interfere with observations.[114] The parasol consisted of a large screen with an aperture in the center, and included a mechanism for adjusting its elevation and angle. The sidereal clock, made by M. Tiede of Berlin, hung on the south face of a pier dedicated to this purpose, which was located just northwest of the meridian circle. [see Figures 4.3 and 4.4] The stone pier provided stability required for the clock to maintain regularity. A similar clock by Tiede hung in the Berlin Observatory and was praised for its ability to keep uniformly accurate time for long periods.[115]

Figure 5.1. Daguerreotype self-portrait of Henry Fitz, Jr. taken in 1839 is among the earliest daguerreotype portraits ever made. (Photographic History Collection, National Museum of American History, Smithsonian Institution)

Chapter 5

Henry Fitz, Jr.:
American Telescope Maker

Until the 1840s, the best telescopes were made in Europe. The telescope had been invented about 1600 and was brought to America around 1650. By 1750, the telescope was commonly found in the libraries of cultivated gentlemen. [see Figure 3.4] David Rittenhouse, a Philadelphia mechanic who began his career as a clock maker, was the first American telescope maker of note. He was followed by Amasa Holcolm, a Massachusetts surveyor, who in 1830 presented to Professor Silliman of Yale College a telescope he had constructed himself. Holcolm was subsequently successful in dominating the American telescope market from 1833 to 1845.

By 1853, when Tappan was shopping for a refracting telescope, a self-trained optician named Henry N. Fitz, Jr. was making telescopes at his shop at 237 Fifth Street in New York City—telescopes that had captured the attention and the confidence of American astronomers.[116] Between 1840 and 1855, Fitz made forty percent of all telescopes sold in the U.S., and eighty percent of all telescopes manufactured in the United States.[117] Fitz's telescopes cost far less than those of European makers, and were deemed to be equal in quality. National pride may have been a further inducement to purchase an American-made telescope, along with the time and money saved by not having to ship a telescope from Europe. Americans were not producing meridian circle telescopes of European quality, though, and so Tappan ordered that telescope from Pistor & Martins of Berlin.

Henry Fitz, Jr. had begun his career as a printer, following his father's trade. He rapidly became skilled, but he did not like the printing trade, so he turned his attention to locksmithing through an apprenticeship spent in the shop of William Day of New York.[118] While saving for a locksmith shop of his own, he pursued his hobby of astronomy. He made his first telescope in 1838—a reflector—and worked as a speculum[119] maker for a time. Upon reading about Daguerre's work in photography, Fitz decided to travel to Europe to learn this process as well as to learn more about optics and optical glass. Daguerre's new process had captured great attention, and the race was on to bring his discovery to America.

As happens with many discoveries, it is not clear to whom the rights to the first American daguerreotype portrait should be attributed. It is commonly held that Samuel F. B. Morse and John W. Draper, both faculty members at NYU, were involved very early, and that they made portraits at about the same time. Daguerre had used his new process to make photographs of landscapes, and had told Morse that the taking of portraits by the daguerreotype process was impossible.[120]

Upon returning to New York in November of 1839, Fitz became involved with daguerreotypes by assisting Wolcott and Johnson in making improvements to a camera: Fitz used his expertise to polish the speculum.[121] Both Morse and Draper made photographic portraits, Morse of his daughter and Draper of his sister.[122] John W. Draper, in his *Scientific Memoirs,* claims that he invented the art of taking photographic portraits.[123] At about the same time, Fitz made a "likeness" of his sister, Susan, which is among the earliest daguerreotype portraits ever made in the United States.[124] Fitz's son, Henry Giles (Harry) Fitz, later gave testimony that he believed Fitz's portrait of his Aunt Susan was the first ever made.[125] The self-portrait Fitz made around 1839 shows his eyes closed, a strategy used to reduce the possibility of a blurred image caused by an inability to remain still during the 20-minute exposure time.[126] [Figure 5.1]

Controversy over the identity of *the* true inventor was commonplace during this era of scientific discovery, and NYU had more than its share of debates. The debate over the daguerreotype portrait was mild compared to disputes over the invention of the electric telegraph. Samuel Finley Breese Morse, a professor at NYU at the time he invented the electric telegraph, fought for years for recognition of his work. Testimonials, such as that provided for Henry Fitz by his son, were the principal basis for the settlement of claims over inventions. Henry Tappan was called upon by Morse in 1868 to write such a letter to verify what Tappan had observed regarding Morse's invention while they were on the faculty together at NYU. Tappan had witnessed Morse's practical demonstration of the electric telegraph at NYU in 1835 with a small group of others. In February of 1868, Tappan wrote to his wife while he was away in Berlin to share with her a transcript of his letter about Morse. The letter captures Tappan's excitement about scientific discovery, and his thrill about being associated with the men responsible for such inventions:

My dear Julia,
I send you a copy of my letter to Morse in reply to his. You can lay them away together. They may both be interesting hereafter. Fulton was a pupil of West's in London as well as Morse. They were not there at the same time. Fulton pursued painting for some time before he became engaged with the steam engine. It is curious that the painter Fulton should be father of Steam Navigation, and indeed also of Railroads, for the propulsion of vessels on water led to the propulsion of locomotives on land, by steam: and the painter Morse the father of the Electro-magnetic Telegraph. What a change these two inventions have made in the

world! How they have carried forward the centuries! And they both owe their inventions to American painters!

You are my dear
    Julia
    HPT[127]

Morse was successful in his lengthy succession of law suits, receiving a large fortune in royalties. He went on to invent the submarine telegraph, later assisting Cyrus W. Field and Peter Cooper in laying the first Atlantic cable.[128] He was active in photography, as well, constructing the apparatus with which he took the first pictures of the sun ever made in the United States. Tappan and Fitz were in the company of some of the most brilliant scientific men of the era.

Henry Fitz's daguerreotype studio, located at 112 Baltimore St. in Baltimore,[129] which he operated from 1840-42, was successful, enabling Fitz to open a telescope and camera business in New York in 1844. Fitz married Julia Ann Wells of Southold, Long Island in June of 1844 after a decade of friendship and correspondence. The marriage served Fitz well: Julia, who was intelligent, literary, artistic and personable, encouraged him to continue building telescopes.[130] Although telescopes were to become the principal commodity of his business, Fitz remained interested in photography.

It was in October of 1845 that Fitz exhibited at the Fair of the American Institute, held annually in New York, a 6-inch refracting telescope. It was modeled after an instrument made by the German telescope masters, Merz and Mahler, in 1840 for the Philadelphia High School Observatory.[131] This was the turning point of Fitz's career as a telescope maker: he was recognized with a gold medal. But, this was to be just the first of a long string of high honors and awards that would be bestowed upon Fitz. His telescope, with a tripod considered to be ingenious, and an achromatic objective that possessed unique qualities of Fitz's own invention, attracted much attention. He opened a telescope shop at 237 Third Street, New York in 1847.

The astronomer, Professor Elias Loomis, praised Fitz telescopes in his book *On the Progress of Astronomy* and in an article for *Harper's New Monthly Magazine,* as had Lieutenant Gillis of the U.S. Coast Survey in his report of his expedition to Chile.[132] Fitz made use of these comments in the text of the advertisement for his business. Professor Hiram Mattison also touted Fitz's telescopes in his book, *A High School Astronomy:*

It is but recently that any good refracting telescopes have been made in this country. The best have formerly been made in Germany and France; but they are now manufactured with success, and to considerable extent . . . His telescopes are perfectly achromatic, and are sold much cheaper than imported ones of the same size and value. . . . Besides patronizing a worthy American optician, [you] will get as good a telescope and much better mounting than by sending abroad, and at far less expense.[133]

Fitz used American flint glass, but he imported Guinand's French crown glass for his lenses,[134] ordering blocks or plates of it in sizes ranging from 2 inches to 24 inches square and about an inch thick. The outsides of these plates were rough, but they were absolutely clear when looking through them edgeways. Fitz used a primitive method for cutting the glass into circular shapes: he bent a sheet metal strip into a hoop, fastened it to a wooden disk with a shaft to spin it rapidly, and brought the strip down onto the glass while an assistant spooned on wet sand. The grains of sand that were rubbed onto the glass helped the metal to slowly cut its way through. Lenses were ground by preparing a cast iron bed in the proper form, and then employing the following process:

> A rod, of the length of the radius corresponding with the curve desired, is secured upon a table by a pivot at one end, when a cutting tool in the other end of the rod is swept across a plate of sheet iron, cutting it in two. This curved edge is then used as a guide for turning a pattern for casting the inner bed. This bed is then placed upon the top of a revolving shaft, and, being covered with wet sand, serves as a cutting surface for grinding down the lens into a corresponding form. . . . After the lenses are reduced to the proper form, they are polished with oxyd of iron in molds of bell-metal, similar to those of iron employed in the grinding operation.[135]

Although Fitz employed women to grind his lenses, Fitz did the more precise figuring and polishing.[136] He kept his process a secret, without recording its details on paper. The special process Fitz had developed for "local zone" polishing corrected any inhomogeneities he spotted in the optical glass by grinding gently with optical rouge using his fingertips,[137] a process so intense it would split open his skin. The warmth of his fingertips helped to accelerate the flow and figure of the glass.[138] Fitz wrote to his father in 1845: "I feel sure I have a way not known to any other of making the corrections, and perfecting an achromatic."[139]

While Fitz never published instructions for making telescopes, he shared some of the details of his process in a letter to James C. Watson, Brünnow's student at the Detroit Observatory, who was attempting to make telescopes himself:

> The pitch is our common black pitch as you see by the enclosed fragments cut from [sic] polisher. I put it upon the tool from a spoon, in the melted state, (say about ⅛ of an inch thick on the face, though less would work, and when the tool is evenly covered, apply [to] the lens, or if you choose to keep on the safest side, apply the tool well wetted to the surface and form it completely with that, before attempting the lens). You can boil the pitch to a proper consistency, or if very soft, mix some resin. The specimen enclosed you can apply to a tool as I often do by merely examining the metal and pressing the pitch upon it.
>
> To turn off the edges and center the lenses you can either attach them to a chuck upon a lathe and hold a hoop of their sheet iron, (or keep the iron fast) only meeting the lens when a projection is to be cut down. We do not often scratch the bare lenses and never protect them.

To obtain the true optical center of your lenses, two methods are adopted generally, with success and ease, viz: to use a perforated mandril upon which the lens is set with pitch and made to form an image, which of course remains at rest while the lens revolves, or to place the lens against a truly and smoothly turned [th--] bearing of lead or pewter, then bringing the reflection to remain at rest while the lens revolves. Another plan is adopted by the Munich Opticians, which is simply measuring the thickness by means of very accurate feeling calipers, and turning the excess off in the lathe by emery laid upon a hoop fixed to the lathe.[140]

Beyond his great talent and tireless effort, two factors proved to be influential in Fitz's success at launching his telescope business. First, he was able to manufacture telescopes at one-quarter to one-half the price of European-made instruments. Second most important to his future career was the attention of amateur astronomer, Lewis M. Rutherfurd, a wealthy New York lawyer and trustee of Columbia College, whose interest in astronomy had been aroused while he was an undergraduate at Williams College. Rutherfurd's order for a 4-inch refractor, and his outspoken confidence in Fitz's abilities, attracted orders from others, first from amateur astronomers such as Robert Van Arsdale of Newark and Jacob Campbell of New York, but also from universities and colleges such as Michigan and Vassar, and observatories such as the Allegheny Observatory in Pittsburgh and the Dudley Observatory in Albany. Rutherfurd continued to be Fitz's patron, ordering numerous telescopes over the years and praising Fitz widely. Their friendship undoubtedly influenced Fitz to build a home in 1863 on 11th Street in New York, near Rutherfurd's.

Fitz's career was in full bloom when his life was cut short in 1863. The exact cause of his premature death is unclear. One account describes a dramatic demise in which the chandelier in Fitz's new house fell on him, resulting in his death a few days later. Another, perhaps more believable, account indicates that Fitz died following a brief illness, just prior to his departure for Europe where he planned to purchase glass for a 24-inch refracting telescope lens. He was also about to apply for patents for a camera with a new form of lens, which indicates that his interest and involvement in photography continued until his death.[141] Fitz's son, Harry, carried on his father's telescope business sporadically over the next twenty years, with the advice and backing of Lewis Rutherfurd. Harry discontinued his telescope work around 1884 to devote full attention to the teaching of drawing, though he continued to make astronomical observations. Harry died in 1939 at the age of ninety-one. Henry Fitz's talents were also carried on by one of his students, John Byrne, who opened a shop in New York at 314 East 21st Street and became a master at making smaller objective lenses, mostly for amateur astronomers who treasured their quality.[142]

The firm Alvan Clark & Sons had taken command of the American telescope market by 1866, when they completed an 18½-inch refractor, the largest objective in the world at that time. It was the Clarks who gradually refigured many of Fitz's lenses because, after Fitz's death, and as the technology for making telescope lenses developed, some of Fitz's lenses came to be seen as deficient. The objective of the Detroit Observatory's Fitz telescope appears to have remained untouched—a testament to its high quality. Today, it stands unique as the largest remaining Henry Fitz lens that is not known to have been refigured.

Tappan placed his order with Henry Fitz in February of 1853. Tappan traveled with Henry Walker from Detroit to New York City, where they met with Fitz and several other scientific gentlemen at the residence of Julia Tappan's father, Colonel John W. Livingston. There, a contract was established whereby Fitz would furnish an achromatic refractor, equatorially mounted, of at least 12 inches clear aperture, to be delivered in New York ready for transportation to Ann Arbor on or before the first day of June 1854. Upon completion, Fitz was to receive $6,150.

Dr. Charles W. Hackley of Columbia College, a former colleague of Tappan at NYU who was dismissed along with Tappan, may have been one of the scientists present that day. Hackley, who had been educated at West Point, went on to become President of Jefferson College in Mississippi. Returning in 1843 to New York as Professor of Mathematics and Astronomy at Columbia, he assumed the Chair of Astronomy alone in 1857. Hackley had been "particularly active in his efforts to establish an astronomical observatory in New York City."[143] He was also active in publishing. His works included a *Treatise on Algebra*, an *Elementary Course in Geometry*, and *Elements of Trigonometry* (with its application to navigation and surveying, nautical and practical astronomy and geodesy). In addition, in 1856 he edited Charles Haslett's *Mechanic's, Machinist's, and Engineer's Practical Book of Reference*, a volume to which Hackley added the *Engineer's Field Book with Formulae for Running* [Railroad] *Lines, Locating Side Tacks, &c.*

It is known that Tappan and Hackley kept in contact after they left NYU. Tappan mentions Hackley in his correspondence with Jasper Cropsey (see Chapter 7), suggesting that Hackley and Cropsey knew each other. Hackley may have introduced Tappan to Cropsey, or suggested that Samuel P. Avery introduce Tappan to Cropsey. Hackley visited Tappan in Ann Arbor in 1856, shortly after the Cropseys' visit, in the company of Dr. Edmund Andrews. Before Andrews resigned from the University of Michigan in June of 1855, he had been Professor of Comparative Anatomy and Demonstrator in Human Anatomy, as well as the Superintendent of Buildings and Grounds at the time the Detroit Observatory was constructed. He went on to a similar faculty position at Rush Medical College in Chicago, becoming one of the original faculty members of the Chicago Medical College three years later, and founder of the Chicago Academy of Sciences.

The reason for Hackley and Andrews' visit to Ann Arbor in early August 1856 was to accompany Tappan to a meeting of the American Association for the Advancement of Education on August 12th in Detroit, for which Tappan was preparing an address called "John Milton on Education."[144] From Detroit, they traveled together to Albany, accompanied by Franz Brünnow, to attend the dedication of the Dudley Observatory on August 28th and the coincident scientific meeting of the American Association for the Advancement of Science, at which Brünnow presented a scientific paper, "Tables of the Asteroid."[145] From Albany, it appears that Tappan and Brünnow traveled to New York City.[146] In a letter written to Henry Fitz while he was making the refracting telescope for the Detroit Observatory, Hackley alerted Fitz of the forthcoming visit, suggesting that "the first item was not good and the other will not be ready for two weeks."[147] This may refer to problems with the telescope Fitz was making for Michigan. Given Hackley's friendship with Fitz, it is quite likely that Hackley had recommended Fitz to Tappan to construct a telescope for Michigan.

The size of an objective lens was a competitive feature of telescopes. When ordering new telescopes, astronomers desired to surpass existing telescopes in size, as a measure of superiority. The telescope for Michigan was to be second only to the great refractors at Cambridge (Harvard) and Pulkova in Russia, both of which had 15-inch objective lenses. It would have a clear aperture of 12⅝ inches and a 17-foot 8-inch focal length. With it were included seven negative and six positive eye pieces, the highest magnifying power being 1,200 times and the lowest 50 times; a ring-micrometer; sun shades; a filar-micrometer of German construction with one vertical wire and three fixed horizontal wires; and a position circle reading by two verniers to single minutes. The achromatic finder was 36 inches in focal length with a 2¼-inch clear aperture. The equatorial mounting was of the Fraunhofer design with a polar axis of 5 inches diameter at the larger bearing, 2½ inches at the smaller end, and 35 inches long. Both axes were made of bell metal, and the hour circle was of brass, 18 inches in diameter, graduated on an inlaid band of silver to single minutes, and read by means of two verniers to single seconds of time. The declination circle was of brass, 20 inches in diameter, graduated on a band of silver to 10 minutes of arc from 0 degrees to 160 degrees, and read by means of two verniers to 10 seconds of arc. To the hour circle was attached a tangent screw for slow motion in declination. There was also clock work connected with the hour circle to give the telescope slow motion corresponding exactly to the diurnal motion of the stars and planets arising from the rotation of the earth on its axis.[148]

The telescope's wooden tube was constructed of pine covered with strips of mahogany veneer, and along each side of the telescope were long, wooden rods that controlled for flexure in the tube. [Figure 5.2 and back cover] At the ends of these rods were attached large brass balls filled with lead which served as counterweights. O. M. Mitchel of the Cincinnati Observatory described his similar Fitz refractor as being so simple of operation that even a one-year-old child could with ease move the telescope by lightly touching these balls with the palm of the hand.[149] This delicacy of motion was achieved using a system of friction rollers and counterpoises.

Figure 5.2. Refracting telescope (circa 1895) was made by Henry Fitz, Jr. in 1857 for the Detroit Observatory. Its 12⅝-inch objective lens made it one of the largest refracting telescopes in the world. (Bentley Historical Library)

President Tappan reported to the Board of Regents in October 1854 that part of the Fitz equatorial was on its way to Ann Arbor, with the remaining part to be brought to Michigan by Fitz, who would then mount the refractor in the Observatory's dome. Most likely, Fitz transported the objective lens to Ann Arbor himself, not trusting its transport by any other means. Fitz's achromatic lenses were particularly delicate because he ground the flint glass so thin.[150] [Figure 5.3] But, Tappan's announcement to the Regents was premature: Fitz was far behind schedule. Brünnow had arrived in Ann Arbor in July 1854, immediately assuming his duties, but the Fitz telescope was not ready. In order to satisfy his contract, and Brünnow's undoubted impatience, Fitz provided a telescope on loan in April 1855. The telescope intended for the Observatory did not arrive until December 1855, nineteen months late. What caused this delay is not known, but after the long wait, the telescope was found to be unacceptable:

> The moulding . . . with the exception of the position circles and the appendages belonging to the tube, was of cast iron, and the experience of a few months having clearly demonstrated its inferiority, rendering the instrument nearly if not entirely useless for making astronomical observations, a new contract

was made with Mr. Fitz by which he agreed to make a new instrument to be mounted wholly in brass and bell metal. For this, when completed, he was to receive the one already in use and the additional amount of $600, making the total cost of the new instrument $6,750.[151]

Fitz's use of cast iron in the construction of his telescopes had been praised by many, in part because the new style mounts sold for considerably less than the original. Fitz's friend, Hiram Mattison, added positive comments in a revision of his *A High School Astronomy*. Mattison wrote in 1853:

> Mr. Fitz has recently made a very valuable improvement in the mounting of tele-
> scopes — one which is not only much superior to the old method, but which costs
> only about one-half as much. This improvement consists in using a single piece
> of cast-iron in the place of several pieces of brass work. It is very simple, secures
> great steadiness to the instrument, and is easily adjusted.[152]

Obviously, Brünnow did not agree. Given Brünnow's exacting standards (see Chapter 6), it is not surprising that he found the new mount deficient.

In November of 1857, Fitz delivered to Michigan an acceptable telescope, which was the third to be mounted in the dome. Ironically, the advantages of American telescopes in terms of decreased delivery time and cost did not prove to be an advantage for Michigan: a satisfactory telescope was three years late in arriving. The object glass of the second telescope must have also been deemed defective because the third telescope provided by Fitz included a new lens.[153] In September 1856, the University had George Saunders make a stone pedestal [pier] on which to mount this new telescope, for which he was paid $115. This confirms that the first two telescopes had been mounted on wooden piers, which would have been a much less stable configuration.

The mount for this third telescope was made by Jonas H. Phelps of Troy, New York, under Fitz's direction.[154] Jonas Phelps was a machinist who specialized in scientific instruments. He had made the mount for the Alvan Clark & Sons 7⅛-inch refracting telescope installed in the Hopkins Observatory at Williams College in 1852,[155] and was later recommended to make the mount for a similar telescope at Jefferson College in 1859.[156]

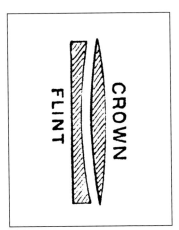

Figure 5.3. Anatomy of a doublet achromatic lens. (*Harper's New Monthly Magazine*, 1874)

Jonas Phelps, who was born in Watervliet, New York in 1809, was working as a bell founder at age 24 in Troy, New York. His business, located on Fulton Street, evolved to include the manufacture of mathematical, civil engineering, and surveying instruments. As business expanded, Phelps moved in 1844 to 319 River Street and took on an apprentice named Lewis E. Gurley. The next year, Gurley's brother, William, who had been the foreman for Oscar Hanks' mathematical instruments business in Troy, became a partner with Phelps. The firm's name changed to Phelps & Gurley, and in 1851 to Phelps & Gurleys when Lewis Gurley became a partner. Their business, manufacturing surveyor's compasses, theodolites, transits, leveling instruments, goniometers, measuring chains and tapes, microscopes, telescopes, drafting instruments and other scientific apparatus, continued to flourish. They moved into larger quarters at 514 Fulton Street, and rebuilt in the same location after a devastating fire in 1860.

Phelps sold his interest in Phelps & Gurleys in 1852, and the name was changed to W. & L. E. Gurley, a firm which gained significant renown.[157] The firm still exists today in its 1860 building on Fulton Street, known for a time as Teledyne Gurley and now as Gurley Precision Instruments, Inc. [Figure 5.4] The firm maintains a small museum of its historical scientific instruments, including some made by Phelps, while the New York State Museum holds in storage

Figure 5.4. W. & L. E. Gurley surveying instruments company (as it appeared in 1997) still operates in Troy, New York. The original Phelps & Gurley company was founded by Jonas Phelps who made the mount for Michigan's Fitz refracting telescope (Patricia S. Whitesell)

an extensive collection of scientific instruments made by Phelps and the Gurleys. Phelps relocated in 1852 to Westport, Connecticut, where he continued to make surveying instruments, and so it was from Westport, just northeast of New York City, that he worked with Henry Fitz in 1857 on the telescope mount for the Detroit Observatory. When Jonas Phelps died in 1865, the Gurleys purchased his finished and partially-finished instruments, including the circular dividing engine Phelps had designed in Westport and which the Gurleys apparently wished to manufacture.[158]

Altogether, three separate Henry Fitz telescopes occupied the dome at the Detroit Observatory: the original 1854 telescope which was rejected; the 1855 loaner; and the 1857 telescope which Brünnow found to be acceptable. The endeavor of repeatedly moving a 17-foot long telescope weighing about a ton in and out of the dome, not to mention hoisting up a 4-ton stone pier, must have been a challenge. The telescope crate would have arrived at Ann Arbor's railroad depot, and then been transported by wagon to the Observatory's grounds, and up the steep hill.

Asaph Hall, Jr., director of the Observatory from 1892-1905, had used a tackle rigged to the frame of the dome's shutter in order to lower the telescope tube and axes so they could be cleaned of accumulated oil and grease.[159] It is speculated that a similar process was used to hoist the three Fitz telescopes into the dome, with the tube entering the building through the front doors of the Main Hall, continuing through the archway into the Inner Hall, and then hoisted up into the dome through a hatch opening in the ceiling. Mathematical calculations have verified that, based on the geometry of the building, this method was feasible.[160] The stone pier was likely lifted into the dome through the floor hatch using the same means.[161] A 2-inch diameter hole exists, drilled through the middle of the stone pier, through which a chain or cable was threaded. It is known that Henry Fitz was reimbursed $14.25 by the Regents for the purchase of "chain for dome" in October of 1856.[162] The timing of the reimbursement suggests that the chain was purchased in order to hoist the loaner telescope into the dome, and probably used again to remove the loaner telescope and install the third.

In 1906, work on the reconfiguration of the Fitz refracting telescope began under the supervision of William J. Hussey, the newly-appointed director of the Observatory. The original wooden tube was removed and replaced with a steel tube painted glossy black. [Figures 5.5 and 10.10] Rumor has it that the original wooden tube is in storage in the basement of a campus dormitory near the Observatory, though searches have [so far] been unsuccessful in locating it. The dew cap from the original wooden tube telescope was retained. The dew cap is a sleeve of metal which fits over the end of the telescope to keep condensation from forming on the lens. It was painted to simulate the wood grain of the original telescope tube, and was never repainted to match the steel tube, so it provides documentation of the color and graining pattern of the original

Figure 5.5. Michigan's Fitz refracting telescope as it appeared after the 1907 replacement of the original wooden tube with one made of steel. (Bentley Historical Library)

wooden tube. Other changes to the telescope included a new 3½-inch finder, a new driving clock, a course circle in right ascension and another in declination, a slow motion and clamp in right ascension and another in declination, a worm and worm wheel, a counterweight arm and weights, and a right ascension circle. All of the work was done in the Observatory Shop by instrument maker Henry J. Colliau under Hussey's direction.[163]

Over the years, the Fitz refractor received several coats of paint, and its general condition deteriorated due to accumulations of grease and grime, exposure to the elements, disuse, and theft of its eyepieces. [Figure 5.6 in centerfold] Nonetheless, it continued to function. The refractor was fully restored in 1997 to its 1907 configuration. [Figure 5.7 in centerfold] The telescope's original Fitz objective lens is particularly notable in that it is unique in the world: it is the oldest surviving, large Henry Fitz objective that has not been refigured by another telescope maker. Astronomers and amateurs interested in the history of astronomy, and in Fitz lenses in particular, eagerly await the University of Michigan's scientific examination of this lens—research that will document its unique characteristics, with the hope of learning more about Fitz's unique optical techniques. A scale model of the Fitz refractor as it appeared in 1857, with its wooden tube and flexure rods, is being created so that visitors to the Detroit Observatory can see the telescope's original appearance and learn about the technology of wooden-tubed telescopes.

Figure 6.1. Franz F. E. Brünnow, first director of the Detroit Observatory and the University of Michigan's first Ph.D. faculty member, was recruited from Berlin. He became Tappan's son-in-law in 1858. (Bentley Historical Library)

Chapter 6

The German Connection: Recruitment of Franz Brünnow

Franz Brünnow (pronounced brew-nuv) was recruited from Germany by President Tappan in 1854 to be the first director of the Detroit Observatory. [Figure 6.1] This appointment brought to the University the first faculty member to hold the Ph.D. degree. It also brought to an American university the German method of astronomy, which was distinguished in its level of mathematical rigor. It would not be long until Ann Arbor became regarded as *the* place to study astronomy, referred to by some as the Ann Arbor School of astronomy. When Cleveland Abbe decided to pursue advanced studies in astronomy after graduating from the Free Academy of New York in 1857, he heard from every astronomer he consulted that, if he could not study in Europe, he must enroll at the University of Michigan, which he did.[164] The students who received instruction from Brünnow, including Abbe and his classmates, Asaph Hall, Sr. and James Craig Watson, went on to lead distinguished careers in astronomy (see Appendix 5).

Brünnow was not the first astronomer to whom Tappan offered the position. The first offer was extended to W. A. Norton of Yale College, who turned it down, as did Benjamin Apthorp Gould of the U.S. Coast Survey. The idea of appointing Brünnow came to Tappan while reading a letter Brünnow wrote regarding the meridian circle telescope. Tappan had engaged Drs. Brünnow and Encke of the Berlin Observatory in Prussia (Germany) to inspect the meridian circle and the Tiede sidereal clock prior to shipment to the United States, in order to ensure that the instruments were properly constructed. In his letter, Brünnow expressed great enthusiasm about the meridian circle, indicating he would envy the astronomer whose good fortune it was to use the equipment. Tappan later reflected "It then occurred to me that his admiration might lead him to accompany it."[165]

Tappan wrote to Brünnow to express the desirability of such an arrangement, and at the same time wrote to several noted astronomers for their opinions. There was enthusiastic support for Brünnow, with the exception of B. A. Gould, who had already turned down the position. Gould held the opinion that foreigners should not be recruited to fill American faculty positions. This prejudice would later appear again in Gould's work at the Dudley Observatory in Albany, this time against Dr. C. H. F. Peters, another German astronomer.[166] However, Tappan believed otherwise, following the advice of the many who spoke in praise of Brünnow. Brünnow's career was being championed by his mentor, Professor Johann F. Encke, of the Berlin Observatory, and by the great scientist, F. H. Alexander von Humboldt.[167] Humboldt later wrote of Brünnow:

> Mr. Brünnow, as everyone knows who is conversant with the literature of modern astronomy, occupies in public opinion a place equally distinguished by the solidity, I ought to add, the sagacity, which reigns in his publications, by the universality of his experience as an observer, by that intimate knowledge of the construction of instruments optical, and of measurement so necessary in the establishment of a great observatory. The supreme direction of an institute worthy of the States which move at the head of the civilization of the New World, cannot be entrusted to more worthy hands. [He is] attached heart and soul, like myself, to the prosperity, the grandeur to the intellectual progress of your noble country . . .[168]

Brünnow accepted the appointment with the strong encouragement of Encke and Humboldt, arriving in Ann Arbor in July 1854. The successful recruitment of Franz Brünnow as the inaugural director of the University of Michigan's Detroit Observatory in 1854 brought to America the German method of astronomy for its first formal introduction into the American curriculum. The significance of Brünnow's contribution to American higher education of European methods and rigor in astronomy has been likened in significance to the contributions of Alexander Agassiz in natural history.

Brünnow was 33 years old when he came to Michigan as the first Professor of Astronomy and Director of the Detroit Observatory. His arrival in Ann Arbor was noteworthy, as captured in a letter written on August 21, 1854 by a student, William D. Anderson:

> We have built an observatory here and have called one of the best astronomers in the world to take charge of it, namely Dr. Brünnow of Berlin Prussia. On his way here he stopped at some of the [?] colleges and caused many of them to envy the University of Michigan its good fortune on obtaining the services of such a man. His name will bring a large procession of students . . .[169]

Dr. Brünnow made it possible for Michigan to advance to the forefront of instruction in astronomy. The scientific curriculum Tappan introduced in 1852-53 included for freshmen the study of algebra, geometry, and history; sophomores concentrated on trigonometry and conic sections, descriptive and analytic geometry, mensuration and navigation, natural history, German, history, and rhetoric; juniors studied natural philosophy, German, drawing, perspective and architecture, calculus, rhetoric, civil engineering, and chemistry; and seniors took civil engineering, chemistry, animal and vegetable physiology, and geology. Interestingly, astronomy was offered through the classical, rather than the scientific, course to third term juniors. What distinguished Michigan from other institutions was Tappan's University course, a newly instituted graduate program which students could enter following receipt of a bachelor's degree. This University course included astronomy. Students who wanted to pursue a course in higher astronomy began in their senior year and continued for four semesters, taking them one year beyond graduation. Brünnow's course in higher astronomy included:

1. Spherical Astronomy and Theory of the Instruments
2. Calculation of the Orbits of Celestial Bodies
3. Numerical Calculus; Theory of Interpolation; Method of the Least Squares
4. Physical Astronomy; Calculation of the Special and General Perturbations of the Heavenly Bodies

This rigorous theoretical and highly technical training, and practical experience through hands-on use of the new meridian circle and refracting telescopes, was not available to students at the other three comparably equipped observatories in the country. Throughout Brünnow's impressive career, he was dedicated to the training of future astronomers.

Brünnow was born Franz Friedrich Ernst Brünnow on November 18, 1821 in Berlin, Prussia (Germany). He was the son of Johann Brünnow, a German privy councillor of state, and Wilhelmine Weppler Brünnow. He was educated at the Friedrich-Wilhelm Gymnasium and the University of Berlin, studying mathematics, astronomy and physics. In 1843, following the preparation of his thesis titled *De Attractione Moleculari*, he received the Ph.D. degree. He then became the director of the private Bilk Observatory at Düsseldorf, Germany in 1847, where he authored an important paper on De Vico's comet and was honored the following year with a Gold Medal from the Amsterdam Academy. He moved to Berlin in 1851 to be Encke's first assistant, replacing Johann Gottfried Galle who was appointed Director of the Bresslau Observatory. Brünnow was said to be the favorite pupil of the renowned astronomer Encke, trained among a distinguished group of young astronomers that included Galle, Bremiker, and D'Arrest. Brünnow was present when the planet Neptune was discovered by Galle on September 23, 1846, a discovery which had been anticipated based on predictions made earlier by Urbain J. LeVerrier of France.

It was at the time he moved to Berlin that Brünnow published his most important work, *Lehrbuch der spärischen Astronomie*, or *Handbook of Spherical Astronomy*, the first edition of which contained a preface by Encke. This text reached its fourth edition, and then Part I was translated into English by Main, the Radcliffe Observer at Oxford, in 1860. Brünnow himself translated the entire text into English in 1865, with subsequent translations published in Spanish, French, Russian, and Italian. This text was held in high esteem by astronomers, and it established Brünnow as an astronomer of great international renown.

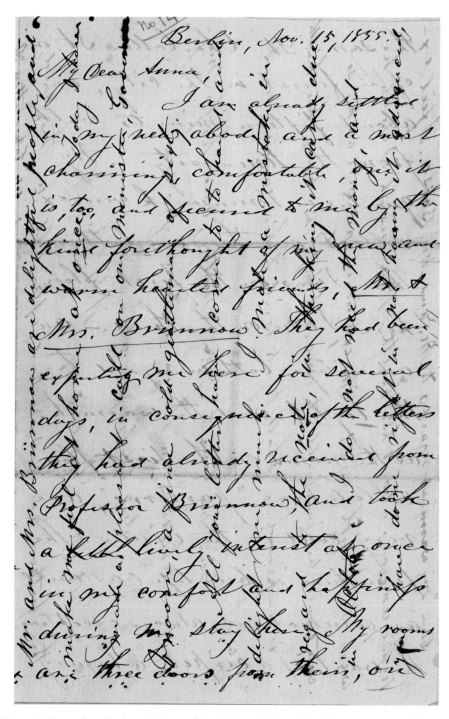

Figure 6.2. Letter from Professor Frieze to his wife from Berlin, where he was visiting with Franz Brünnow's family, was written in the crossed style to save paper and postage. (Bentley Historical Library)

As a special tribute, Brünnow dedicated the English translation of his *Handbook on Spherical Astronomy* to George P. Williams, who was professor of mathematics and science at Michigan, the first faculty member appointed at the University, in 1841, and the instructor in astronomy prior to Brünnow's arrival. Brünnow was particularly influenced by Professor Williams, holding him in high regard. The esteem Brünnow held for Williams was quite a tribute considering the rigorously high standards Brünnow held. Professor of Latin, Henry S. Frieze, who later became Acting President of the University, described Williams in a memoir written about President Tappan:

> He was one of the best of men; thoroughly sincere, absolutely free from all obliquity. His genial kindness, his pleasantry, his ever ready repartee, made him the favorite among all the professors of two generations. But he possessed also traits which are not so generally known. He was a man of earnest piety and of strong religious convictions, and he carried into his life the religion he professed. Perhaps also it is not generally understood that he was something more than an ordinary teacher. He was a profound scientist . . .[170]

Brünnow was comparably respected because he possessed similar traits, characterized by high standards and ideals, extremely hard work, and amazing perseverance.

It is possible to gain an impression of Franz Brünnow's family upbringing through the letters of Dr. Frieze who wrote regularly to his wife, Anna, during his visit to Berlin in 1855. Frieze was authorized by President Tappan in June 1855 to undertake a tour of Europe for the purpose of visiting various seats of learning. Arriving in Berlin in November 1855, Frieze was greeted by Mr. and Mrs. Brünnow, who had been alerted by their son of Dr. Frieze's impending visit. The Brünnows had engaged rooms for Frieze at 94 Charlotten St., three doors down from their own residence. Frieze wrote to Anna, "I am already settled into my new abode, and a most charming and comfortable one it is, too, and secured to me by the kind forethought of my new and warm hearted friends, Mr. and Mrs. Brünnow."[171] In order to conserve paper and reduce the cost of postage, Frieze wrote in the crossed style that was common at the time. This style of letter was first read left to right, then bottom to top, starting at the lower left corner. [Figure 6.2]

The Brünnows were a large and close family, consisting of three separate families that assembled regularly, never missing an opportunity to celebrate birthdays. On the occasion of Mrs. Brünnow's birthday on February 20, 1856, the third Brünnow family birthday event Frieze attended during the first three months of his visit in Berlin, the others being for Professor Brünnow and his father, Franz was remembered through a "curious anecdote" related to a previous celebration of the same occasion:

> The gist of it was that when [Franz] returned to the observatory that night, he rang the wrong bell, and found some difficulty in adjusting the instruments for making observations, though none in seeing <u>double</u> stars.[172]

The study of double stars was an obsession of Franz Brünnow's, with particular attention paid to double-star 85 Pegasi, the measure of which he calculated in 1870.[173]

Physically, Brünnow was described as being "of a modest demeanor, with dark long hair, and exceedingly lustrous dark eyes."[174] He was particularly personable, though quiet, and was an exceptionally talented musician, joining with Professor Frieze, Mrs. Andrew D. White, and Rebecca Tappan in a chamber music quartet that performed in amateur concerts. Brünnow and Rebecca were both pianists, Brünnow being quite an accomplished student of Beethoven. The proceeds of public concerts held by their music club supported the University's purchase in 1862 of a statue of "Nydia," Blind Flower Girl of Pompeii, sculpted by Randolph Rogers, an internationally-known artist who had spent his boyhood in Ann Arbor.[175] [Figure 6.3] Acquisitions such as this supported Tappan's desire to establish museum collections.

Figure 6.3. Statue of Nydia (still extant) by Randolph Rogers, a sculptor of international renown who spent his youth in Ann Arbor. Franz Brünnow and Rebecca Tappan helped purchase the sculpture by performing in amateur concerts. (Bentley Historical Library)

Some found it difficult to understand Brünnow's "broken English" due to his thick German accent, but his writing was eloquent, and flawless in either language. His writings portray a very well-educated man—a scholar who was thoughtful and confident in his abilities and knowledge, and a true gentleman in every regard. He commanded respect based on all of these qualities. Although he spent only nine years in America, he left his mark in the history of American astronomy, and was considered to be one of the best of a small number of astronomers in America in the 19th century.

Brünnow took an interest in Tappan's daughter, Rebecca, or "Barbie" as she was affectionately called. In January 1856, Professor Boise wrote to Dr. Frieze in Berlin that "Dr. Brünnow and Miss Rebecca seem to be on excellent terms."[176] They were married on March 15, 1857. Of the marriage, Tappan wrote to Jasper Cropsey "Two beings never seemed happier."[177] Brünnow requested and was granted an 8-month leave of absence, and the couple sailed for Europe on March 21st aboard the steamer Hermann. They stopped in Southampton for a few hours, then traveled on to Bremen, finally reaching Berlin where they spent the summer with Brünnow's family. They returned to Ann Arbor the following November, arriving in time to supervise the installation of the Fitz refracting telescope in the dome of the Observatory.

Rebecca Tappan was energetic and adventurous, her father's cherished child. To Brünnow, she must have seemed worldly compared to other Michiganders, having traveled to Europe with her father when she was fifteen years old—even acting as interpreter when Tappan met at the Sorbonne with his mentor, the great philosopher M. Victor Cousin, because Tappan was not proficient in the French language. Tappan wrote that Cousin "seemed touched with the enthusiasm of a young heart."[178] Cousin presented Barbie with one of his books, a gift to which she responded with an enthusiastic letter, written in flawless French, described by Tappan as representative of her "girlish simplicity."

Although Tappan did not approve of having women attend universities because he considered women to be a distracting influence, it is clear that Rebecca had received, and continued to pursue independently, an excellent primary education. She was fluent in several languages, including French and German, and in her later years, she published her translations from German to English of poetry by Joseph Victor von Scheffel, including *The Trumpeter of Säkkingen* (1877) and *Mountain Psalms* (1882).[179] Her son Rudolph wrote in 1883 to Professor E. P. Evans at the University, with whom the Tappans had kept up a regular correspondence, making reference to his mother's translation:

> Many thanks for your kind article in the "Allgemine," which has given my Mother great satisfaction. I trust that attention will be called to the matter in England; it is a shame that my Mother's translation should be so utterly ignored. It is, however, I believe, the publisher's Chapman's fault.[180]

Franz Brünnow knew well the importance of publishing. While at Michigan, he started the University's first scholarly journal, *Astronomical Notices*, the purpose of which was, primarily, to publish the discoveries and research findings made at the Detroit Observatory. However, contributions from other notable astronomers were also published. O. M. Mitchel of the Cincinnati Observatory had, in 1846, started a journal called *The Sidereal Messenger*. The European equivalent was Berlin's *Astronomische Nachrichten*, published continuously from 1821 to the present day except for a 3-year lapse during World War II.

Research occupied a great deal of Brünnow's time and attention. His work involved complex and time-consuming computations, often focused on the detailed study of the motion of asteroids. He published several such studies, including "The General Perturbations and Elliptical Elements of Vesta,"[181] which appeared in *Astronomical Notices*, and *Tables of Victoria*,[182] published by the Regents in 1858. His publications brought to Michigan a reputation for excellence, and his ephemerides provided practical astronomers with valuable information to assist their understanding of the placement of celestial bodies (see Appendix 2).

Besides being a renowned astronomer, Franz Brünnow made contributions to the advancement of higher education in America, in cooperation with Tappan. It had long been Tappan's desire to advance his philosophy of higher education through the creation of a national university that included an academy for science and arts. New York was the obvious center for such an institution as it was the center of intellectual activity in America in the 19th century, and Peter Cooper[183] was motivated to make a financial investment in the idea. In January 1856, while Tappan was at Michigan, he wrote to Samuel B. Ruggles[184] of New York letting him know that he had just sent to Peter Cooper a document describing a plan for the creation of what Tappan proposed be called "The University & Academy of Sciences & Arts in the City of New York."

The plan was the inspiration of Tappan, but it had been enhanced by Brünnow based on his knowledge of the university system employed in Berlin:

> The union of the Academy with the University—the Academy being comprised of the University professors with the power to associate others with them, is a new idea for which I am indebted to Dr. B[rünnow]. It is an improvement upon Berlin & Paris—the Academy in the first, and the Institute in the second being distinct from the University in organization, although composed in part of the same men.[185]

Tappan's correspondence continued through the winter, with letters to the Mayor of New York, to Mr. Astor,[186] and to Samuel B. Ruggles. Finally, in April 1856, a response came that Peter Cooper had contributed $100,000 toward new professorships as a start. However, Cooper wished to have the institution named after himself, and preferred the name Cooper Scientific Institute. Tappan expressed his desire that they stick to the word "Union," which must have been proposed as a revision to Tappan's original name for the institution.

Thus, the Cooper Union for the Advancement of Science and Art was created in the City of New York, thanks in large part to the ideas and persistence of Tappan and Brünnow. The Cooper Union building was erected on Eighth Street between Third and Fourth Avenues, and included instruction in science, mathematics, geology, chemistry, physics and astronomy, art, design, mechanical drawing, telegraphy, mechanics, engraving, and oratory.

Franz Brünnow resigned his position as Professor of Astronomy and Director of the Detroit Observatory shortly after the Regents dismissed President Tappan from his position at the University on June 25, 1863 [see Chapter 13]. Brünnow had little choice, given the situation. On June 13, 1863 Tappan had written to the Honorable H. Barney, Collector of New York, requesting the renewal of Brünnow's passport as he intended to sail alone for Germany on July 4th, the timing of which suggests that Brünnow had made plans to visit his family in Berlin over the summer vacation, before he learned that Tappan was dismissed.[187] Tappan and his family followed Brünnow to Europe in the fall of 1863, after numerous, impassioned attempts to restore Tappan to his position failed to succeed.[188]

Although the reasons for Brünnow's resignation seem clear, James Watson and others presented a different explanation for Brünnow's departure. In a letter to George P. Bond, who was director of Harvard's observatory, Watson wrote that "Dr. Brünnow resigned his connection with the University on account of continued ill-health which his physician attributed to disagreement of our climate with his physical organization, and which he feared might result, if it had not already resulted in organic disease of the liver."[189] But, it is clear from Bond's response that he knew the truth. He replied, "I had supposed that Dr. Brünnow's retirement was mainly attributable to recent changes in the administration of the affairs of the University. I hope that his return to Germany may prove an effectual remedy for his complaint and restore to him that sympathy & encouragement in scientific pursuits which he must have found sadly wanting on this side of the Atlantic."[190]

Brünnow went on to become, in 1865, Astronomer Royal of Ireland, the Andrews Professor of Astronomy in the University of Dublin, and Director of the Dunsink Observatory. [Figure 6.4] At Dunsink, he continued his research on stellar parallax, which he published in his *Astronomical Observations* (1870) and *Researches Made at Dunsink* (1873). This work was described as "lucid in arrangement, clear in exposition, every part of the work given in sufficient detail to enable anyone, with but little trouble, to follow the various steps of the reductions. . . . clearly the high order of accuracy attainable by a skilled observer . . ."[191]

Brünnow's appointment at Dunsink was an ideal situation, except that Henry Tappan found Ireland disagreeable in terms of its culture,[192] which limited the number and duration of visits by the Tappans. Dunsink, which means "Hill of Hills," is located on a lofty prominence four miles outside of Dublin, and is still a functioning observatory today. Tappan described Dunsink while visiting there in 1866:

We have here a fine view of the surrounding country bounded by hills & mountains, and the harbor of Dublin. The air is pure & bracing—never very hot never very cold—a wonderfully equable temperature. The grounds are ample— 25 acres—beautiful gardens—beautiful trees and flowers—a large fine house— with carriage horse stables &c. The Doctor has two cows which afford him plenty of Milk and butter—he has two carriages, a covered one & an open one—he has a farmer, a gardener and a coachman. We drive about daily & find the best of society. . . . Dr. Brünnow has absolute control of the Observatory & is hard at work as usual.[193]

In 1869, Brünnow was elected a Fellow of the Royal Astronomical Society, and in 1871, he expanded and updated with John Stubbs of Trinity College the classic text titled *Brinkley's Astronomy*. Brünnow remained at Dunsink until 1874 when his failing eyesight forced his resignation. He retired to Basle, and in 1880 to Vevey, Switzerland to be with the Tappans, settling finally in Heidelberg, Germany in 1889 after the Tappans' deaths in 1881 and 1884. His poor eyesight precluded any scientific work. Mostly, Brünnow occupied himself with his other passion—music—for which he had great talent and fondness. In a moment of reflection, it is said that he once remarked that had he not pursued astronomy, he ought to have devoted himself entirely to music.[194]

Figure 6.4. Dunsink Observatory near Dublin, Ireland where Franz Brünnow was director and Astronomer Royal after leaving Ann Arbor in 1863. (Dunsink Observatory)

Figure 6.5. Rudolph Brünnow, taken in Ann Arbor in 1914. (Bentley Historical Library)

Figure 6.6. Rudolph's son, Eric Brünnow, taken in Ann Arbor in 1914. (Bentley Historical Library)

Brünnow's death on August 20, 1891 at age sixty-nine was unexpected, though he had been seriously ill several months earlier. He had been making preparations for a trip to Switzerland when he suddenly became ill. Rebecca Brünnow had long suffered from ill health. The severity of her symptoms seemed to vary depending on the climate of their particular locale. In 1879, Tappan wrote to his friend Mr. Murphy in Ann Arbor "she goes on about the same. Not worse—better perhaps; but the lump is there and the pain and weariness & the morphine injections twice a day. What the denouement will be God only knows."[195] She died in 1893.

The Brünnows had one son, Rudolph Ernst, born February 7, 1858. [Figure 6.5] He was almost constantly under the guidance of his grandfather, and theirs was a very close relationship. At the dedication of a University memorial to President Tappan in 1914, Rudolph, accompanied by his son, Eric,[196] [Figure 6.6] gave an impassioned speech in Ann Arbor about Tappan's influence on his life:

> Up to my twenty-third year I had the privilege of being almost constantly under the influence of his inspiring personality; and his precept and example had a preponderant part in the formation of my character and tastes. The lofty idealism that pervaded his whole being was to me a beacon-light which guided me through many a dark hour of doubt and uncertainty; and his broad-minded sympathy with every manifestation of human thought and effort has always stood before me as the goal to which all true intellectual culture should aspire. . . . the romantic interest that he wove around the scenes of the Biblical tale left an impression on my youthful mind that has never been effaced and led me on, as I grew older, to an ever increasing love for the Ancient East.[197]

Specifically, Rudolph traced the interests that led to his scholarly career to "the wonderfully impressive way in which [Tappan] used to relate the story of Joseph and his sojourn in Egypt" at age five in the President's House. Evidence exists that Rudolph's interest in Egypt continued throughout his youth, and greatly influenced his career direction. [Figure 6.7] In 1872, while at Dunsink Observatory outside Dublin, the Brünnows were visited by Dr. Charles Piazzi Smyth, Astronomer Royal for Scotland. In his diary, Smyth describes the visit:

25 July 1872

Dr. Brünnow meets us with two-horse carriage at the station; particularly warm and friendly greeting . . . Introduced to Madame Brünnow and to the son, greatly taken with Δ*, and to Father-in-law, a fine aquiline-countenanced old American once a Presbyterian preacher, since President of Michigan College in U.S. He also well pleased with Δ talk.[198]

*Smyth used the symbol Δ as shorthand for pyramid.

Rudolph was educated at Saint Columbia's College near Dublin; Trinity College, Dublin; and the Universities of Basel, Tübingen, and Strassburg, receiving his doctorate from Strassburg in 1882. He became a distinguished scholar and professor of Semitic Philology at the University of Heidelberg from 1884-1904, and finished his career at Princeton University as chair of Semitics from 1910-17. Being a man of independent means,[199] he was able to pursue his research in between his faculty appointments, undertaking expeditions to southern Syria in 1895, 1897, and 1898, which resulted in the publication of his Magnum Opus, *Die Provincia Arabia* in three volumes. He later published, after painstaking research, a *Classified List of Cuneiform Ideographs*, which was considered an indispensable handbook for Assyriologists. He was expert in many Semitic languages, including Assyrian, Arabic, Hebrew, Aramaic, Syriac, and Phoenician.

Rudolph was dedicated and accomplished in his academic specialty, receiving the prized Drexel gold medal for distinguished service to archeology. But, his interests were wide-ranging and varied, which was attributed to his father's broad interests in mathematics, physical sciences, and music. Personally, Rudolph was described as having "a fine appreciation of music . . . modest in character, generous, public spirited and very lovable."[200]

On June 1, 1894, Rudolph married Marguerite Beckwith of New York, a granddaughter of Edwards Pierrepont, former Minister to the Court of St. James's.[201] They had five children: Eric, Katherine,[202] Marguerite,[203] Hildegarde, and Richard. Rudolph was a professor in the University of Heidelberg, but was largely an independent scholar. He and his family later resided at Châlet Beauval in Vevey, Switzerland, where they lived happily [see Figures 13.1 and 13.2]. Grief stricken when his wife died in 1907, he moved the family to Princeton, New Jersey to carry out his wife's wishes to have the children educated in the United States. He purchased a magnificent summer home at Mount Desert in Maine, a lifestyle which was undoubtedly made possible through the inheritance of Mrs. Tappan's wealth and the sale of Châlet Beauval.

Tragedy struck again in 1916 when Eric died of infantile paralysis. Rudolph's life was nominal from that point, having lost his spirit through the deaths of his wife and eldest son. Rudolph died in 1917 and was mourned not only by the academic community but by the residents of Mount Desert who had delighted in "his quiet charm, his devotion to local interests, especially to the formation of paths and woodland trails."[204] Unfortunately, Rudolph's estate, valued in excess of a half million dollars, dwindled to $45,000 by an unfortunate coincidence of timing: war with Germany had been declared just one week previous and bank stock that had been purchased at peak prices suffered a significant loss. His home in Princeton, New Jersey, purchased for $200,000, was sold for only $73,000, and his charming cottage in Mount Desert was sold at a loss of $170,000.[205] What was left was willed to the four surviving children, all of Garrison, New York.

Figure 6.7. Henry Philip Tappan, taken in Dublin circa 1872 while visiting the Brünnows at Dunsink Observatory. (Bentley Historical Library)

Figure 7.3. Jasper F. Cropsey portrait by Edward L. Mooney. (Newington-Cropsey Foundation)

Chapter 7

Jasper F. Cropsey:
Hudson River School Painter

In 1855, Tappan commissioned Jasper Francis Cropsey, a landscape painter of the Hudson River School of artists, to paint a landscape of the campus and the newly completed Detroit Observatory.[206] [Figures 7.1 and 7.2 in centerfold, and the frontispiece] It appears that Tappan met the Cropseys through Samuel P. Avery, the New York art collector. Tappan and Avery met with Cropsey in New York in August 1855,[207] and Cropsey, accompanied by his wife, Maria Cooley, traveled to Ann Arbor in September of that year to create what would be the earliest recorded image of the Observatory. The painting is done in tondo, which means in a circular shape.

The Sturges family of New York, whose children Tappan had escorted to Europe and with whom Tappan had made arrangements to receive the meridian circle at New York harbor for transfer on to Ann Arbor, were patrons of Cropsey. So were other famous New Yorkers, including Cyrus W. Field,[208] who laid the trans-Atlantic telegraph cable, the Astors, and the Morgans.[209] The Tappans were to become patrons as well.

The Cropseys became very close friends of the Tappans. Even though Tappan had little time for correspondence due to demands on his time, being "overrun with business [with] scarcely a moment to [himself]"—refusing to be a regular correspondent with anyone—he was active in his correspondence with the Cropseys after their visit, and he made it clear to them that he wrote more often to them than to anyone else.[210] Tappan's letters to the Cropseys, to whom he usually wrote separately, were particularly full of warmth and intimacy.[211] Cropsey sent the Tappans some engravings in January of 1856, and for "Barbie" Tappan he sent a sketch of Hyde Park and a "charming little graph."[212] As a demonstration of friendship, Tappan offered that Mrs. Cropsey should add a volume of Shakespeare to the list of five books he had recommended for her enjoyment, and that she should obtain a copy at Appletons in New York and charge it to his account. Appletons was the publisher of Tappan's *Elements of Logic*, a copy of which he gave Cropsey as a gift, with the handwritten inscription "Mr. J. F. Cropsey from his friend the Author, Jan, 1856."[213]

Cropsey was born on February 18, 1823 on the family farm in Rossville (Staten Island), New York, the oldest of eight children. [Figure 7.3] Cropsey had an interest in art from the time he was a young boy, drawing sketches of landscapes and buildings in the margins of his school books. He took an interest in architecture, creating an elaborate model of a house that was exacting in its level of fine detail. This won him a diploma at a fair of the Mechanics Institute of the City of New York in 1837. Shortly thereafter, he began his career as an architect, working as an apprentice to New York architect Joseph Trench. Cropsey had a rigorous apprenticeship, which included the study of geometry and draftsmanship, and after a short time, he was reviewing all drawings that left Trench's office. During this time, Cropsey studied watercolor under Edward Maury, and won a diploma from the American Institute for an architectural drawing of the Temple of Minerva, Parthenon Restored. He then took up oil painting, studying under various artists associated with the National Academy of Design, and was elected an Associate of the National Academy in 1844 at the young age of twenty-two.

In the meanwhile, Cropsey had ventured out on his own in 1842, opening his own office at 72 Chambers Street in New York. His first commission was a Greek Revival church at New Dorp, Staten Island. Only one other set of plans survives from this early period of Cropsey's architectural career, so it is tempting to speculate that, due to Cropsey's architectural talents, he may have assisted in the design of the Detroit Observatory (see Chapter 8), there is no evidence to support that he did.[214]

Cropsey was encouraged as a painter by artists William T. Ranney and William Sidney Mount, and when he exhibited his first painting in 1843 at the National Academy of Design, it was well received. Cropsey conducted independent studies of linear perspective and the theory of light and shadow, which had applications in both architecture and art. His dual careers of architecture and painting benefited one another: the detailed drawing he learned as an architect exhibited the fine details and the precise perspective seen in his paintings. The detailed nature of his art is evident in his painting of the Detroit Observatory: each wooden slat of the exterior shutters is distinct, and even the cables that lift the Meridian Room roof hatch can be seen. Cropsey married Maria Cooley of West Milford, New Jersey in 1847 and left immediately for a European tour where he was influenced by artists such as Christopher Cranch and Thomas Hicks. [Figure 7.4]

In 1856, the Tappans commissioned Cropsey, whom his friends called Frank, to do a painting for them of Genevieve, inspired by a favorite poem by Samuel Taylor Coleridge, titled "Love," in which Genevieve is the central character.[215] Coleridge's "Dissertation on the Science of Method" had influenced Tappan as a philosopher,[216] and Tappan enjoyed Coleridge as a poet as well. When Tappan's extensive library collection was sold in 1876, Tappan requested that two books be held back: the family *Bible* and Coleridge's *Aids to Reflection* in which he had made notes. Cropsey's painting of Genevieve was received by the Tappans in Ann Arbor on March 11, 1856 and hung in the President's House to the left side of the fireplace in the front room.[217] The entire family was thrilled with the painting, but Brünnow, in particular, was charmed by it. In the spirit of a true astronomer, he studied the shadows Cropsey had painted, questioning whether the shadows were correctly portrayed. Tappan wrote to Cropsey:

> [Brünnow] says the full moon is rising out of the ocean, and of course the sun is setting opposite. The shadow which falls against the tower and broken wall is the shadow of some object out of sight, but then he asks, why does not the shadow fall also upon the mound where Genevieve and her lover stand? With this he is puzzled. Mrs. Tappan and I say the figures ought not to be in shadow. But he says, the same object which throws a shadow upon the tower and at the foot of it, must fall upon the figures also.[218]

When the Cropseys and their daughters, Minnie (born 1850) and Jennie (born 1852),[219] moved to England in the summer of 1856, Tappan was deeply disappointed by the loss of his friends. "Plague on this changeful world" Tappan wrote to Maria Cropsey. "I have just gotten to know you & Frank and to love you, and now you are going to leave me."[220] Tappan's concern for their

Figure 7.4. Maria Cooley Cropsey was a close friend of Tappan, and they corresponded regularly. (Newington-Cropsey Foundation)

safe voyage reveals his personal fear of icebergs—a fear he confesses was exceeded by nothing else. In March, Tappan urged that the Cropseys delay their voyage for at least a month until the icebergs melted. A news clipping titled "How Steamships are Lost" was included with the letter to punctuate the point. It appears that Tappan's advice was followed, for the Cropseys arrived in London in late July, though the ship was delayed by traveling far off course, having lost its bearings during a storm, taking a full month to arrive at its destination instead of the usual twelve days. While sailing, Cropsey painted a picture of his family in their steamship stateroom, about which, with friendly confidence in the artist's abilities, and with characteristic wit, Tappan queried "What if you should accomplish such a triumph of art as to produce sea sickness in the beholder!"[221] [Figure 7.5] It was the exception for Cropsey to paint human forms, explained in this case by the lack of any other subject during the month-long voyage.

He built up his career over the next decade, and was honored in 1862 by appointment by Queen Victoria to the American Commission of the London Exposition. In 1866, he helped to establish the American Watercolor Society, and he was honored with a medal for Excellence in Oils at the Philadelphia

Figure 7.5. Jasper F. Cropsey painted his family in their ship stateroom en route to England in 1856. (Newington-Cropsey Foundation)

Centennial Exposition in 1876. By this time, his paintings were selling for prices up to $2,500 (approximately $25,000 in 1998 dollars). Toward the end of his career, he designed several platforms and waiting stations for the Sixth Avenue Elevated Railway in New York City, and two churches in Staten Island.

After suffering a severe stroke in 1893, Cropsey died in 1900 at his residence, called Ever Rest, in Hastings-on-Hudson, New York, which has been preserved by the Newington-Cropsey Foundation. The Foundation has constructed a new building to house and display Cropsey's paintings, architectural drawings, and other memorabilia. When Cropsey died, he was in the process of painting a portrait of his wife; the unfinished painting now hangs in his preserved studio at Ever Rest.

Cropsey's 1855 painting of the Detroit Observatory was engraved for the University by R. Leggett and used as University letterhead, in the annual *Catalogue*, and elsewhere, as was Cropsey's landscape painting of the campus. [see frontispiece and Figures 7.1 and 7.2 in centerfold] Cropsey's Ann Arbor paintings became widely recognized, and admired, and are still extensively used in publications today. Andrew D. White, a former professor of history at Michigan who went on to become a U.S. Senator, U.S. Minister and Ambassador to Germany, and later to Russia, president of the American delegation to the Hague Peace Conference in 1899, and the first president of Cornell University, acquired the two paintings in 1863, later presenting them to the University in 1890.[222] White came into possession of the paintings at the time President Tappan was removed from the University by the Regents. It is likely that most of Tappan's belongings, and things associated with him, were sold or destroyed at this time, a suspicion supported by an obvious absence in the University's archives of the bulk of the Tappan and Brünnow papers.[223]

While Cropsey was in Ann Arbor, and continuing on through his return trip to New York by way of Lake Erie, the St. Lawrence, Montreal, Quebec, and Lake Champlain, he kept a sketchbook in which was included a sketch of the President's residence and some small sketches and notes that helped him remember details for the campus landscape painting. [see Figure 1.2] The sketchbook was later donated to the University by Professor Bruce M. Donaldson of the University's Department of Fine Arts, who obtained the scrapbook from Mrs. Max Winkler, widow of the late Professor of German who had it in his possession for many years.[224]

Today, Cropsey's paintings are highly valued and prominently displayed in museums, at the Newington-Cropsey Foundation, and in private collections across the country. The University of Michigan's Bentley Historical Library holds the campus landscape, Observatory painting, and scrapbook, and the Museum of Art holds a painting done shortly before Cropsey's death and two pencil sketches.

Figure 8.1. Professor Richard Harrison Bull (circa 1857) of the University of the City of New York designed the Detroit Observatory. (New York University Archives)

Chapter 8

The Observatory's Architect

Upon returning from New York in March 1853, where he saw Tappan depart for Europe to purchase astronomical instruments, Henry N. Walker reported to the Regents by letter that a Mr. George Bird of New York had been engaged by Tappan to furnish a plan for the Observatory and, if necessary, to superintend the construction for a fee not to exceed $300.[225] Researchers have for years been puzzled by this reference. Mr. Bird does not appear in any relevant records for the City of New York. The error may have been one of transcription when the contents of Walker's letter were copied by hand into the Regents' minutes book, or Walker may have confused the name with someone else he met in New York, or perhaps Tappan may have changed his mind after Mr. Bird was engaged.[226]

Although an explanation regarding Mr. Bird may never be found, all speculation about the identity of the Observatory's architect can now be put aside. A venture inside the mind of Henry Tappan is what finally unlocked the elusive identity of the man who designed the Observatory. Tappan was well-connected in New York, and he had a history of turning to his friends and colleagues for referrals and assistance. Several clues were uncovered that narrowed and focused the search toward Tappan's colleagues at NYU, where he had been on the faculty from 1832 to 1838. In the archives of Hamilton College, in Clinton, New York, was found a newspaper clipping from an 1858 edition of the *Albany Argus* which had been saved by Professor of Astronomy, Dr. C. H. F. Peters. The article, which laid out the details of the construction of the Detroit Observatory, identified a Professor Buhl of New York as the architect.[227] Frederick and Christian Buhl of Detroit, also mentioned in the article, were donors to the Detroit Observatory, so it seemed possible they had suggested to Tappan that he engage a relative of theirs to be the architect. This lead proved false. The *History of Washtenaw County*, the *Albany Argus* article reprinted in a local Ann Arbor newspaper, and Charles Kendall Adams' *Historical Sketch of the University of Michigan* (1876) all indicated that a Professor Bull of New York was the architect, suggesting that the reference to Professor Buhl was a typographical error. This proved to be the case.

Turning to Theodore F. Jones' *History of New York University 1832:1932*, it was here that Professor Richard Harrison Bull's identity was established. [Figure 8.1] Dr. Bull was appointed in 1853 as professor of engineering and assistant in mathematics at NYU. The position had been empty since a retirement in 1833. The revival of civil engineering was done in response to the suggestion that NYU establish relations with the City, which was growing at a rapid pace. Bull's revived School of Civil Engineering was combined with Architecture, with the architecture curriculum under the direction of Professor Thomas Seir Cummings, who was a colleague of Professor Samuel F. B. Morse (painter and inventor of the electric telegraph) of the University's School of Design. Professors Bull and Cummings taught mathematics, surveying, architecture and construction practices, including courses in Land Surveying and Leveling; Astronomical Observations and Calculations for determining Latitude and Longitude; Science of Masonry and Carpentry; Principles of Architecture, with elements of Design, Construction, and Estimates; Construction of Railroads; and Railroad Management.[228]

Dr. Bull had the unique combination of astronomical and civil engineering knowledge required to design an observatory, a structure of great complexity and exacting specifications. Bull had access to many individuals with whom he could consult on various aspects of the project—perhaps Jasper Cropsey who was trained in architecture; or Charles Hackley, a counterpart at Columbia University; or Thomas Seir Cummings, Bull's faculty partner in the Civil Engineering and Architecture department and founder of the American Academy of Design; or architect A. J. Davis who designed an observatory (never built) for NYU [see Figure 8.3]; or architects A. J. Downing and Calvert Vaux who drew up the initial plans for the Dudley Observatory [see Figure 8.4]. It is not known who, if anyone, may have assisted Bull, but it is clear that there were a limited number of individuals as well-informed as Bull in his knowledge of astronomy and astronomical instruments, and of civil engineering, who could successfully design an astronomical observatory. It is likely that Dr. Bull designed the internal structure of the observatory building, and that the exterior bracketed embellishments and Doric portico were designed locally given that the style is consistent with other University and local buildings of that era.

Tappan, no doubt, received advice that the Observatory should be designed by someone knowledgeable about astronomy, and Charles Hackley may have recommended Professor Bull. [see Chapter 5] Other astronomers, such as O. M. Mitchel of the Cincinnati and Dudley Observatories, if consulted, certainly would have recommended that someone knowledgeable in astronomy design the building. For the Dudley Observatory in Albany, Mitchel had rejected a plan prepared by professional architects Downing and Vaux, instead enlisting a scientist to do the design, though elements of the architects' plan were retained. [see Chapter 8] It is not known whether Professor Bull traveled to Ann Arbor to assist Professor Silas Douglass in superintending the construction of the Observatory, but there is no doubt that Bull was responsible for its design. Unfortunately, Bull's papers did not survive.

Richard Harrison Bull was born in New York City in 1817, the son of Benjamin and Eliza Bull. He was a member of NYU's class of 1839. After graduation, he studied at the Union Theological Seminary until 1843. He then taught at the University Grammar School until 1847, when he became the secretary and actuary of the Eagle life insurance company. While he was a professor at NYU, he was the secretary and president of the New York Savings Bank, located at Eighth Avenue and Fourteenth Street, an enterprise he built up from its infancy, and which afforded him substantial wealth.

Professor Bull devoted much time and energy to astronomical research. He had a private observatory located at his residence, first at 350 W. 4th and later at New York's prestigious 34 Gramercy Park, and also at his country house at New Hamburg on the Hudson River. "Professor Bull was the first mathematician to put into practice the idea of obtaining the true time by the sun."[229] He used his observatories to furnish exact astronomical time for the New York Central Railroad, an activity he continued until the Western Union erected its timeball.[230] He also held a contract to provide the time to the New York Central and other railroad lines for over twenty years.[231]

Bull's residence at the time of his death in 1892 was an apartment at 34 Gramercy Park East. Its luxurious features were a measure of his affluence. The building, New York City's first cooperative apartment, was erected in 1883 by a group of investors led by Samuel B. Ruggles, the innovative city planner with whom Henry P. Tappan had worked to establish the Cooper Union. Gramercy Park was a prestigious area described as "a crossroads of literature, art, entrepreneurship, philanthropy, invention, political action and social change." The nine-storey, red terra cotta and brick structure housed elegant apartments and featured polished marble floors with tile borders, mahogany, brass, stained glass, and a beautiful mosaic entryway (still extant).[232]

Professor Bull died on February 1, 1892, one month short of his 75th birthday. He was survived by his wife, two sons, Charles and J. Edgar, who were lawyers in the City, and a daughter, Mrs. Charles B. Tooker.[233] According to Bull's death certificate, cirrhosis of the liver, hematemesis, and asthemia were the causes. In his memory, Bull's family commissioned Maitland Armstrong, a well-known artist, to prepare a memorial window in stained glass at the First Presbyterian Church of New York. The window exists today as a lasting tribute to Bull's life, and in particular, to his many years of service as an elder of the Church. Bull is buried at Greenwood cemetery near Prospect Park in Brooklyn.

Dr. Bull was dedicated to the Presbyterian Church, and he avidly pursued the study of theology. At the time of his death, he was preparing to publish an exhaustive treatise on the first book of Genesis, which he had been working on for ten years. Specifically, the book was a study of "the six days of the creation, in which he attempted to prove, through the application of mathematical and

astronomical reasoning, that Moses, by Divine dispensation or otherwise, was cognizant of the laws of mathematics and astronomy when he wrote that part of the Scripture, and that the writing is in harmony with those laws."[234] He had been assisted in this endeavor by a number of well-known theologians. No record can be found that the book was ever published, and its disposition is unknown. During his later years at NYU, Professor Bull took on some odd personality traits that were the target of mischief and teasing by students. He grew a long, white beard that, with his bald head, gave him the appearance of a monastic sage.[235] [Figure 8.2]

Figure 8.2. Dr. Richard H. Bull in old age. (New York University Archives)

Professor Bull had excellent qualifications, yet many observatories were designed by individuals who were not familiar with the unique scientific requirements of an observatory structure. Conflict between the need for scientific exactness versus architectural design was a common problem associated with the construction of observatories, and the issue lends some perspective to the process followed in the design and construction of the Detroit Observatory.

Some famous architects designed observatories, though their designs were not completely successful due to their lack of technical, scientific knowledge. Alexander Jackson Davis, the famous New York architect who had been engaged to prepare the original design for the University of Michigan campus (though it was never executed in favor of building four modest professors' houses), designed an observatory that was to be built at 12th Street and 5th Avenue, which is only a few blocks north of New York University. [Figure 8.3] Like Davis' plan for the University of Michigan, his New York City observatory was never built. The lighthouse style of Davis' observatory is not characteristic of observatory designs: its height and small profile were likely designed to fit the constraints of the site in the urban area for which it was planned.

In 1853, another famous pair of architects, Andrew Jackson Downing and Calvert Vaux, prepared the designs and plans for the Dudley Observatory. However, the director, O. M. Mitchel, was not satisfied with the plans, not due to "any deficiency on the part of the Distinguished Architects" but because

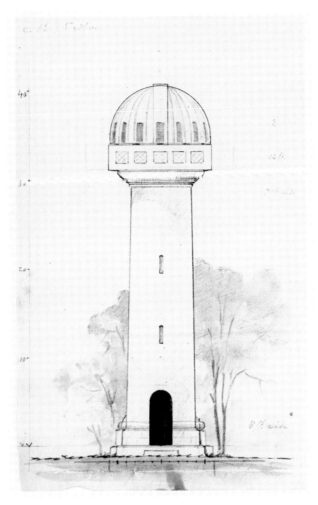

Figure 8.3. Observatory designed by Alexander Jackson Davis for a site at 12th Street and 5th Avenue in New York City, but never built. (Metropolitan Museum of Art, Harris Brisbane Dick Fund, 1924. [24.66.1588])

they did not address the specific scientific needs of an observatory. Mitchel explained ". . . [it] is a simple consequence of the fact, that we were unable to compare our views and wishes by conference and consultation."[236] Initially, in 1852, Mitchel had attempted to accommodate the idea of adapting the Downing/Vaux plan to meet scientific requirements, but his correspondence with the Executive Committee makes clear that Downing did not understand the structural and geographic orientation requirements necessary to accommodate the telescopes. In 1853, Mitchel had Professor George R. Perkins[237] draw up sketches of a plan based on Mitchel's technical needs. Of Perkins' design, Mitchel wrote to the Dudley's Executive Committee, ". . . I think it not deficient in architectural beauty."[238]

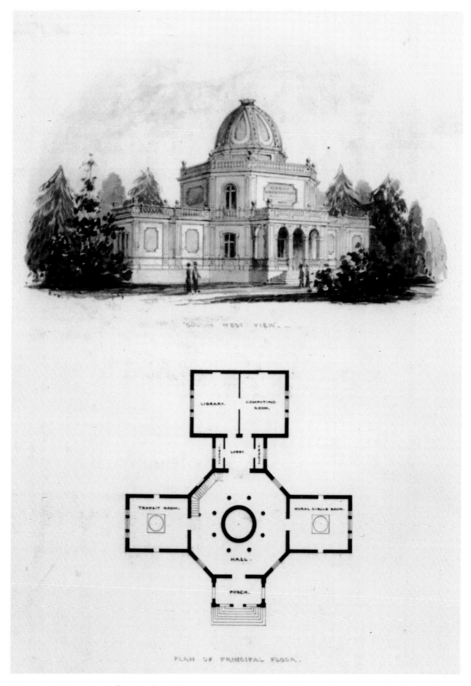

Figure 8.4. Design for the Dudley Observatory by Andrew Jackson Downing and Calvert Vaux. (Dudley Observatory Archives)

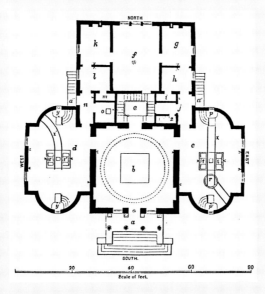

Figure 8.5. Dudley Observatory (no longer extant) following its completion in 1856. Design modifications made by Prof. George Perkins can be identified through comparison with Figure 8.4. (Dudley Observatory Archives; *Annals of the Dudley Observatory, Vol. I,* 1866)

It appears that Perkins' sketches were approved and then turned over to the firm of Woollett & Ogden, architects from Albany, who prepared final drawings.[239] However, certain elements of the Downing/Vaux design were retained, as can be seen in the original watercolor plan prepared by Downing and Vaux. [Figure 8.4] The arched colonnade of four Tuscan columns is identical to the Downing/Vaux design, but a massive, triangular pediment surmounting the columns was added. Despite Mitchel and Perkins' efforts to plan for the scientific requirements of the telescopes, it was necessary to significantly alter the building after construction. Semi-circular projections were added to the front of both wings in order to accommodate the distance required between the north and south collimating telescope piers.[240] [Figure 8.5] In an 1856 newspaper article, these unplanned alterations were described as ". . . varying the outlines of the building, [thereby] add[ing] greatly to its beauty and picturesqueness."[241]

The design process for the Dudley Observatory is one example of the collaborative effort that was needed for the design of observatories. The complex meshing of technical and architectural design elements made it difficult, if not impossible, for the design to be done by a single individual. Although no architectural or construction drawings exist from the 1850s for the Detroit Observatory, documents from the Hamilton College Observatory[242] at Clinton, New York provide a close look at the process involved in designing and constructing an observatory. [Figure 8.6] Although the Hamilton College Observatory closely resembles the Detroit Observatory in architectural style,

Figure 8.6. Hamilton College Observatory (no longer extant), Clinton, New York, constructed in 1854, shown here with its two octagonal additions. (Hamilton College Library)

and was built at the same time, it was Mr. Azel J. Lathrop of nearby Utica, New York who prepared the plans.[243] No complete drawings are extant, but detailed written instructions exist, titled "Specifications of Materials and Work Required for Building an Astronomical Observatory at Clinton Oneida Co. N. York." These specifications, which make reference to the existence of drawings, go into great detail on the excavation of the site and drains; the dimensions of the building and its wings; the foundation, telescope piers, chimney, plaster, and windows. For example:

> Windows in Stone Walls~~~3 windows in foundation 2½ feet wide and 2 feet high with cut stone, caps to sills also two ventilators with common stone for caps & sills and wood frames with cast-iron guard on out side [sic] also two openings in walls from main building to wings for a man to pass with ease.[244]

Two versions of the specifications are extant: a draft contains revisions, and the other, written in much neater handwriting, reflects the changes. Small sketches of windows are provided, offering four different styles from which to choose, one of which was selected and can be seen in photographs of the completed building. Separate sheets were prepared to detail the painting and glazing for the building, with this building to be painted and sanded with two coats, freestone color, except for the back of the building which was just painted—perhaps to save on costs. Bids for this work were obtained from four painters, all coming in fairly close in price. At a later date, octagonal additions were made to the building [see Figure 8.6] for which comparable specifications were drawn up by Frederick Brooks, Builder, included with which was a more detailed cost estimate:

> The octagon to be built with first quality brick the wall to be one foot thick, from seven to eight feet high—eight feet if required for the door into the room. Build a leanto connecting with brick building also with present building-size rooms and dimentions [sic] according to plan drawn by Dr. Peters . . .[245]

The perusal of these fascinating documents, and a glimpse at a few examples of observatory design and construction at other institutions, brings back to life projects that were undertaken nearly 150 years ago, making it easier to imagine the process that went into the design of the Detroit Observatory.

Figure 9.1. View of the Detroit Observatory (faintly visible on the horizon to the right of center) taken in 1919 from Cedar Bend Drive located across the Huron River to the northeast. (Bentley Historical Library)

Chapter 9

Construction of the Detroit Observatory

In March 1853, following Tappan's departure for Europe to purchase a telescope and astronomical clock, Henry Walker returned from New York and requested that the Regents identify a site on the University grounds on which to construct the Observatory. Regent Farnsworth, a donor to the Observatory, established a committee to select a site and to "exercise such control over the erection of the Observatory as may be proper . . ."[246] The Committee included Regent Farnsworth, Henry N. Walker, and Professor Silas H. Douglass. A warrant for $500 was authorized upon which they could draw for expenses. To supplement this, Walker, who had not yet had a chance to collect any of the subscriptions pledged for the Observatory, authorized the University to draw on a $1,000 draft against his account for expenses.[247]

On July 27th, the Regents toured the grounds identified by the Committee as a potential site for the Observatory, but having no quorum present, took no action. At the next meeting of the Regents in November, the land deal was finalized, with Judge Kingsley authorized to purchase land from Sidney and Rachel Benham[248] for $340. Samuel Pettibone[249] surveyed the land for a $3 fee. The site the Committee had selected was a 4-acre section located on a hill a half-mile northeast of the campus, with a commanding view of the Huron River. [Figure 9.1] Local land politics may have contributed to the selection of a site located so far from the campus, but the hill's open vista and distance from the interference of lights, smoke, and the intrusion of other buildings on the skyline were benefits to astronomical observations. Yet, the site proved to be unpopular. Its half-mile distance from campus, and the necessity for travel back and forth, caused the astronomers much fatigue. Before the road was constructed, travel along the dark path after an evening of observations through the telescopes—or even travel during daylight hours—must have been a challenge. [Figure 9.2] Complicating the route was the presence of the "Cat-Hole," a natural hollow between hills which was located in a line from the Observatory to central campus. [Figures 9.3 and 3.10] Water would collect in this area at depths ranging from one to ten feet, which made the site a favorite for pranks, such as dunking. It is said that the 1870 statue of Benjamin Franklin that once graced

Figure 9.2. Rear view of the Detroit Observatory circa 1887 showing out buildings and Students' Observatory (second structure from right). The path leads to a missing picket in the fence where students squeezed through on their shortcut route from central campus. (Bentley Historical Library)

the lawn of the original Law Building may have ended up at the bottom of the Cat-Hole. Legend has it that the Cat-Hole was named for the dead cats often seen floating in the water. Another possible origin to consider is an abbreviated version of Catholepistemiad, which was the original name given to the University in 1817, though the name was quickly dropped.

Tappan had wanted the Observatory constructed in the center of the University grounds,[250] which would have placed it in the middle of what is known today as the Diag. Tappan commented retrospectively:

> I am responsible for everything but [the Observatory's] location. This was determined while I was absent in Europe, without my advice or concurrence. It has proved an inconvenient location, and has caused much fatigue to the astronomer. But, the site is beautiful; and the prejudice which selected one of our highest hills may be pardoned, when one can hardly think of an Observatory without recalling that "Tuscan Artist," Galileo, who, "at evening, from the top of Fiesole, first turned the telescope toward the heavens."[251]

To facilitate access to the site, additional land was acquired from Julia and Philetus Stark for the purpose of constructing a road to the Observatory, with

Silas Douglass and E. W. Stark serving as legal witnesses. The new road, named Observatory Street, was to be four rods wide, beginning at Geddes Road and meeting with Ann Street at the northeast corner of the Observatory property by means of an eastward extension of Ann St. With thoughtful foresight, provision was made in the deed to ensure against any possible encroachment or smoke stack that might interfere with telescope observations:

> . . . we do hereby bargain, sell and convey . . . forever a clean sweep for a telescope for the distance of forty rods from the observatory about to be erected . . . on or above the level of the telescope to be hung in such observatory at the height of twenty five feet or more above the surface of the ground at the observatory, without obstruction by any building of that height or by the smoke of any chimney or smoke pipe within fifteen [sic] below such level of the telescope . . .[252]

With the site selected, building commenced while Tappan was in Europe in 1853 purchasing the astronomical instruments, and the work continued into 1854. The Board of Regents visited the completed Observatory in June of 1855, apparently waiting until Brünnow had the meridian circle in operation. Professor Douglass was appointed to superintend the construction, as he had done for other University buildings, including the South College building, the Medical College, and the Chemical Laboratory.[253] Ground was broken on August 5, 1853 to prepare the site for its stone foundation.[254] What an undertaking this must have been. Observatory Hill, as it was sometimes called, was situated 165 feet above the Huron River. The absence of roads to the site undoubtedly complicated the construction: Observatory Street was not created until 1855. Teams of horses, provided by E. G. Wildt, H. M. Henion, A. Burns, E. Arms, and J. W. Brooks at a total cost of $63.50, would have been used to loosen the soil so that the building site could be leveled, and tons of soil would have been hauled away in wagons from the excavation laborers performed, using shovels as their primary tools. Stones for the foundation would have been hauled to the site in wagons, then carefully cut and fitted together by mason P. Enright, who was paid $20.80. P. Stark provided some stone for the job at a cost of $8.25, and some wood for $9.06. Lumber cost $141.33, provided by H. Moffatt. The joinery work was done by D. E. Wines. Nails were provided by Rulb & Ducharme and H. W. Welles for $21.49, and F & J Hinchman supplied the

Figure 9.3. Newspaper sketch of the "Cat-Hole" with the Detroit Observatory visible in the distance. See Figure 3.10 for a bird's eye view. (Bentley Historical Library)

paints and oils for $87.25. There were many laborers, including George, Dave and J. G. Allmendinger, Patrick Klover, J. F. Fleg, A. Scrocter, J. Fischer, and George Ruff, paid $83.

Joseph H. Vance, the University's Steward, traveled on several occasions for business related to the Observatory. His first trip took place in October 1853, at a cost of $25, which indicates he went a great distance—perhaps to New York to consult with Professor Bull, the building's designer. His subsequent three trips were less costly, charged at $2.20 each, an amount which may have taken him to Detroit and back, perhaps to consult with Henry Walker. Cabinetmaker D. Sperry[255] provided furniture for $13 and George McCollum provided a pump for the well at a cost of $7.50. C. D. Goodrich,[256] who owned the local hardware store, provided pipe, stove, etc., for the Observatory in April of 1856 at a cost of $37.95.[257] Although no chimney appears in Jasper Cropsey's 1855 painting of the Observatory, a chimney is visible in an etching of the building which appeared in an 1859 issue of *Frank Leslie's Illustrated Newspaper.*[258] This evidence, along with the 1856 date for the purchase of the stove, helps to document that a stove was an early afterthought. The telescope rooms were purposely unheated because heat would have interfered with telescope observations. No doubt, that first Michigan winter Brünnow spent in the frigid building, given the milder climate to which he was accustomed in Europe, prompted his request for the stove.

Fortunately, the Observatory never fell victim to fire, as happened to many other early campus buildings. However, during the 1997-98 restoration, evidence of a fire was discovered in the ceiling of the west wing. A three-foot diameter section of the wooden lath and the horsehair plaster keyed into this lathe was found to be seriously charred, although the damage did not extend all the way through the plaster, so that no evidence of the damage could be seen on the interior side of the ceiling. Perhaps a cinder started a fire in the lath as the plaster was being installed, and the fire burned itself out for lack of oxygen.

The exterior finish of the building has long been a source of wonder, prompting much speculation among architects and historians as to why Jasper Cropsey's 1855 painting of the Observatory depicts the exterior stucco as pinkish in color. [see Figure 7.2 in centerfold] The building is constructed of three courses of solid brick covered with stucco that is scored to resemble individual blocks of stone. In the earliest known photograph of the Observatory, taken by T. D. Tooker circa 1858, each scored block appears distinct from the others based on its coloration. [see cover or Figure 14.1] It stands to reason that researchers have long believed the building's stucco was a pinkish color, and that this would account for the pink color used in the Cropsey painting. Others have speculated that Cropsey painted the building a color to his pleasing. But, this is not correct. The explanation is simple, yet it has been overlooked for decades.

In fact, the Observatory appears pink in the Cropsey painting because, in 1855 when the painting was completed, the Observatory's brick exterior had not yet been stuccoed. Close examination of the Cropsey painting, particularly at the southwest corner of the building (it faces due north), reveals the texture of individual bricks. The Cropsey painting is executed with a level of fine detail that rules out any possibility the artist would have used a solid color to depict the stucco treatment, or that he would diverge from the true color. The stucco was highly distinctive, with alternating scored blocks shaded in tones to simulate individual blocks of granite, and Cropsey, with his acute attention to detail, would not have passed up the opportunity to express such exquisite elements of the facade. The building likely remained plain brick until sufficient funds were available and the brick and mortar had cured. Then, the simple building was "dressed up" with stucco to make it appear more impressive.

It was common practice to cover brick structures with stucco in order to give a building a more elegant, finished appearance. The White House in Washington is constructed in this way, and then whitewashed. Alexander Jackson Davis, the architect who prepared the original plan for the University of Michigan campus, designed a mansion in Brooklyn, New York in 1853 for Edwin C. Litchfield, a donor to the Detroit Observatory, which is finished in a faux-stone stucco treatment similar to that used for the Observatory.[259] [Figure 9.4] The J. H. Lund house on Pontiac Trail in Ann Arbor was stuccoed with a

Figure 9.4. Grace Hill, E. C. Litchfield's Brooklyn, NY mansion designed by Alexander Jackson Davis, is now called Litchfield Villa. It had the same stucco treatment as used for the Detroit Observatory. Litchfield was a donor to both the Detroit and Hamilton College Observatories. (League for the Restoration and Preservation of Litchfield Villa)

mixture that contained skimmed milk, and local lore held that a similar mixture was used on the early University buildings.[260]

All speculation about how this stucco finish was achieved can now be put aside. *Dr. Chase's Recipes; or Information for Everybody,* an Ann Arbor-based publication of international acclaim authored by A. W. Chase, provides the recipe used to stucco many of the University buildings, including the Detroit Observatory, the four professors' houses, and the North and South College buildings, as well as other local buildings, such as the Lund house. The fascinating recipe in its full detail follows:

> First make up as much mortar as you need for the job, with good common lime; using only ¾ or four-fifths, at most, as much lime as needed for common work—the other fourth or fifth is to be water lime; and not to be put in only as used. The sand must be coarse and free from loam or dirt.
>
> To prepare the white and colored washes, run off common lime enough with hot water, to make a white-wash to go over the whole job. This white-wash is to be colored the tint desired for the work. Be sure to make color-wash enough at one time, or you will find it hard to get the shades alike; saving a little of the white-wash without color, to pencil the seams, and also for specking, as mentioned below. The colors used are lamp-black, Spanish brown, or Venetian-red, as preferred, and these are cut or dissolved in whisky; then putting into the white-wash to suit.
>
> When these washes are all prepared, wet up as much of the mortar as can be put on in 20 to 40 minutes, and mix in the fourth or fifth of cement, and put on as fast as possible; first wetting the wall very wet with water. Some cement will set in 20 and some in 40 to 50 minutes. When you see the time necessary for the kind you are using, act accordingly, and only mix the cement into as much mortar as your help will put on before it sets; beginning at the top of the wall with your scaffolding and working down, which prevents too much specking from the colors. Have a man follow right after with a float, keeping the stucco very wet while floating down level and smooth; and the longer it is floated and wet the better will be the job. Even after it is floated down well, keep a man wetting it with a brush until you get the whole line on, as the water-lime must be kept quite wet for some considerable time, to set properly. Heed this caution, and if water never gets in behind the plastering from bad cornice or leaky roofs, it will never peel off. When this line of scaffolding is plastered, take out enough of the color-wash, running it through a sieve, and go over the plastering; lamp-black alone gives it a bluish-slate color; if a little of the brown is added with the black, it will be a little reddish, and if the red is used without the brown, it will be quite red. I prefer sufficient of the black only to make a gray stone color. A brown, however, looks exceedingly well. If you choose, you can make one-half of the color-wash darker than the other—having laid it off into blocks resembling stone, by means of a straight-edge, and a piece of board about half an inch thick, paint every other block with the darker wash to represent different shades of stone. Some of our best buildings are done this way and look well.

Then to give it a granite appearance, take a small paintbrush and dip it into the white-wash, saved for this purpose; strike it across a hammer-handle so as to throw the specks from the brush upon the wall, then the same with black and red. Pencil the seams with the white-wash, which gives it the appearance of mortar, as in real stone-work.

Now you are ready to move down the scaffold, and go over the same thing as before. After the colors have been dissolved with spirits, they can be reduced with water, or what is better for them and the color-wash also, is skimmed-milk; and where milk is plenty, it ought to be used in place of water, for white-wash or color-washes, as it helps to resist the weather and prevents the colors from fading . . . Speck quite freely with the white, then about half as much with the black, and then rather free again with the red. The proportion of lime, probably, should not exceed one, to six or seven of sand.[261]

Professor Douglass, who superintended the construction of the Observatory, used a similar stucco treatment for the home he built at 502 East Huron Street in Ann Arbor. It is described in *Dr. Chase's Recipes; or Information for Everybody*, as being "the prettiest color of any [building] in the city." It was done in an imitation of "Free Stone," made with "lamp-black, yellow ochre, and a larger proportion of Spanish brown."[262] The house is still extant, serving as an annex of the Baptist Church, although the original stucco treatment was painted over long ago, just as it was at the Observatory. When this special faux stone stucco finish became soiled over time, the decision to paint over the stucco, rather than to attempt to clean it or recreate it, necessarily prevailed because the intensive labor and technical expertise required for restoration was undoubtedly prohibitive. The Observatory's original stucco was painted over approximately 30 years after the building was erected.[263] Because the Observatory's stucco has been repaired numerous times, primarily where additions have been added and removed, it was not possible to return the building to its original faux stone appearance. Therefore, it was restored in 1997 to its first paint color, a light khaki grey.

All of the exterior trim for E. C. Litchfield's Brooklyn, New York mansion, had been "sanded" to be harmonious with the stuccoed finish. Sanding involved mixing sand with the last two coats of paint and applying it with a wire brush.[264] This technique was also used on the exterior stucco of the Hamilton College Observatory in Upstate New York. However, this method was not employed for the Detroit Observatory: its wood trim and cornice were painted off-white, and the window sash were wood-grained. Wood graining is a faux finish used to dress up common wood to make it look like oak or other hardwoods. An amber or pink base coat is applied under a wood-colored glaze that is patted when wet with a long, natural bristle brush, allowed to dry, and then hand painted to simulate the look of natural wood-grain. At the Detroit Observatory, the window sash, interior woodwork, bookcases in the west wing, dome platform, and stairs in the east wing were wood-grained to simulate natural oak.

The ornamental wooden brackets at the Observatory's cornice add the characteristic detailing of the bracketed style, while the columns and pilasters at the front porch are Greek Revival, a combination of architectural styles typical of the University's early buildings, such as the President's House. Pediments placed above the exterior windows on the ground floor, which deflect water away from the window frame, are made of cast iron, originally painted black. The sills below each window are limestone, as is the water table stone which encircles the entire building in a band located just above the building's stone foundation. These stone embellishments, and the stone used at the front porch, are detailed with textured bush hammering surrounded by grooves chiseled at the edge, a design that is also present on the stone piers for the collimating and refracting telescopes.

The structural design of an observatory is different from all other structures. [Figure 9.5] The main building, which is 32 feet square and 24 feet high, is situated around, but completely detached from, a massive central pier that supports the refracting telescope. The central pier, the purpose of which is to minimize vibration in the telescope mounted at its top, is constructed of solid brick masonry, then stuccoed and scored to resemble individual blocks of stone. To accentuate them, the scored lines in the pier's blue-grey stucco are striped in black, a condition that was revealed during the 1997-98 restoration when the entire pier was scraped of its many layers of paint, an effort that paid off in a

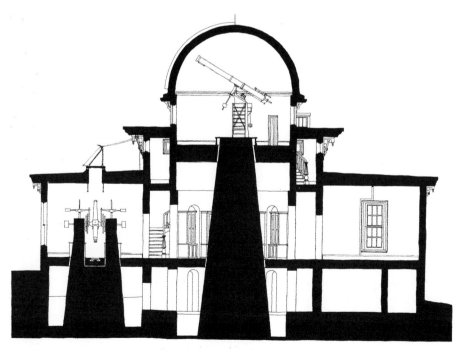

Figure 9.5. Cross section of the Detroit Observatory shows the solid brick telescope piers. (Quinn Evans/Architects)

Figure 9.6. Telescope pier for the 37½-inch reflecting telescope located in the Detroit Observatory's 1908 addition was displayed in Chicago's 1908 Labor Day Parade. The manufacturer considerably exaggerated the pier's weight. (Bentley Historical Library)

stunning result. The pier is submerged fifteen feet below grade, extending up through the center of the Main Hall about forty feet. It is twenty feet in diameter at its base, shaped like the frustum of a cone, tapering to ten feet at the top. It is capped with a circular slab of limestone ten inches thick, on which is mounted a limestone pier nine feet tall and weighing over four tons. To this pier is attached a metal bed plate, to which is secured the refracting telescope.

Hauling the massive slabs of stone from the quarry at Sandusky, Ohio to the Observatory site must have been quite an undertaking. The stone would have been placed on a boat and shipped to Detroit where transfer would have been made to the railroad. Then, once in Ann Arbor, the stone would have been hauled by wagon from the train station up the hill to the Observatory site. In 1908, when addition was constructed and attached to the east wing of the original Observatory structure, a pier was needed for the new 37½-inch reflecting telescope. The iron pier was hauled by a team of 8 horses, an endeavor deemed worthy of display in the Labor Day Parade in Chicago as advertising for the Charles F. Elmes Engineering Works, who made the pier.[265] [Figure 9.6]

A hemispherical dome 23 feet in diameter covers the refracting telescope, with a shutter that opens to expose the telescope to the sky. [see cover or Figure 14.1 and Figures 9.7, 9.13, 9.21, 10.3, 10.7] Originally, this shutter slid on small rollers up and all the way over the other side of the dome, but this method proved so cumbersome that the shutter was replaced in 1890 by a Warner & Swasey model that opened to the side by pulling on a cable. The

Figure 9.7. Detroit Observatory dome circa 1960 showing the original tin roof, catwalks between building additions, and windvane. Lens of the Fitz refracting telescope is visible through the open shutter. (Detroit Observatory Collection)

dome rotates with ease by pulling on a thick, continuous rope. [Figure 9.8] The fact that the dome retains its manual means of operation is noteworthy, because most observatories of the era have had the domes converted to electrical power. The railroad-style wheels on which the dome rotates are not original equip-
ment. Originally, the dome rotated on five cannon balls which ran in grooves in a cast-iron base attached on the wall below and at the base of the dome. [Figure 9.9] The cannon balls design failed, though, because no provision was made to maintain distance between the individual balls: they would eventually cluster together, making it impossible to rotate the dome. The slots that are present on the interior of the Detroit Observatory's dome are said to have been used to insert a board to use as a lever to hoist up the dome so that the cannon balls could be repositioned a distance from one another. [see Figure 5.2] In 1890, the dome was reconfigured by removing the cannon balls and replacing them with railroad-style wheels. [see Figure 5.2] At this same time, exterior brackets were placed under the lip of the dome to provide additional support. [see Figure 9.7]

Figure 9.8. Detroit Observatory's dome is rotated by pulling on a continuous rope. (Patricia S. Whitesell)

Figure 9.9. Dome room of Hamilton College Observatory showing columns that supported the dome and cannon balls (seen above and to the right of the left-hand wooden column) on which the dome rotated. Refractor is by Spenser & Eaton of Canastota, NY (Hamilton College Library)

The use of cannon balls had for decades been the dominant method used to rotate domes, having been devised by William C. Bond of the Harvard College observatory in 1815. The architect Bond had engaged to build Harvard's observatory, Isaiah Rogers, fancied himself an inventor. He had come up with a system for rotating domes using chain rollers, but Bond protested against using the design, convinced it would fail. As the architect for the project, Rogers had his way, but Bond was right: the new system failed. The chain rollers were replaced by cannon balls. Rogers was well known for the design of banks, theaters, and hotels, but he was inexperienced in the design of scientific structures.[266] The Harvard observatory retains on display one of Bond's cannon balls, removed when the dome was later reconfigured to rotate on wheels.

The walls of the Detroit Observatory's dome room are solid brick in a circular form that supports the dome above. The dome was originally supported by wooden columns, but these were determined to be unsafe and were replaced in 1865. Regent Edward C. Walker wrote to President Haven ". . . the wooden columns that support the dome are being taken out and new brick supports from

Figure 9.10. Painter Gary Ontko applies strips of cotton canvas (1997) to the dome's wood sheathing, replicating the original method used to finish the interior of the Detroit Observatory's dome. (Lloyd Baldwin)

the very bottom are being built."[267] The columns probably looked similar to those of the Hamilton College Observatory. [see Figure 9.9] The forces exerted upon the columns were not all vertical: torque would have been created by the rotation of the dome, to which the columns would have offered inadequate resistance, rendering them unsafe only 11 years after construction. A second course of bricks was added to the exterior dome wall for added rigidity. [see Figure 9.7]

The hemispherical dome was constructed of structural timbers spaced at regular intervals that converged at the top of the dome in a complex geometric configuration. These timbers were covered on the exterior with tongue-and-groove boards fitted together with amazing skill. Each board was scored widthwise with a shallow kerf cut every five inches, which allowed the board to bend slightly in order to conform to the shape of the dome. Individual tin plates, with a thin lead coating, formed the exterior armor. Inside, the dome's structural timbers were covered with boards only ⅛ of an inch thick and about 7 inches wide, sheathing the dome's concave interior. The thin boards were then covered with strips of cotton canvas, about 7 inches wide, starting at the bottom of the dome and winding around the perimeter with a 1-inch overlap all the way to the top. The canvas was covered with strips of wallpaper, similarly sized, and overlapped to cover the canvas seams. The small, blue flower pattern on the wallpaper was applied facing toward the canvas so the color would bleed into the canvas rather than through the paint that was then brushed onto the wallpaper. This intricate assembly of layers provided a durable, lightweight,

Figure 7.1. Landscape of the University of Michigan campus painted by Jasper F. Cropsey in 1855 shows the campus from the east. The President's House is second from the left. The other buildings are no longer extant. (Bentley Historical Library)

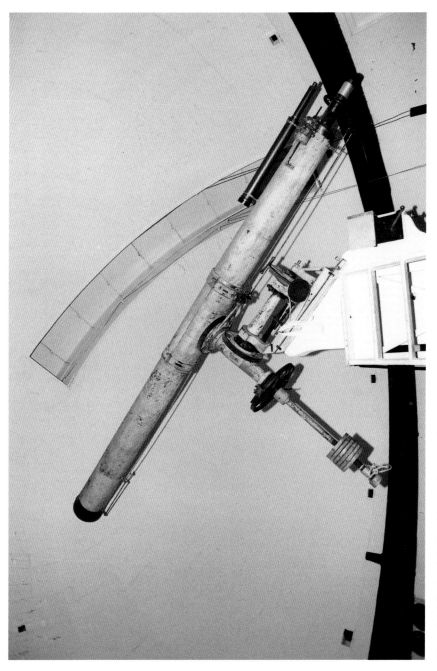

Figure 5.6. Michigan's 12⅝-inch Fitz refracting telescope prior to its 1997 restoration. (Patricia S. Whitesell)

Figure 5.7. Michigan's 12⅝-inch Fitz refracting telescope after its 1997 restoration. (Patrick Seitzer)

Figure 7.2. Jasper F. Cropsey's 1855 painting of the Detroit Observatory. The Observatory appears pink because, at the time of Cropsey's visit to Ann Arbor, the brick building's exterior stucco had not yet been applied. See Figure 14.1 for a view of the building after the stucco was applied. (Bentley Historical Library)

watertight dome with a smoothly-finished interior. During the 1997-98 restoration, the original canvas, which was in fragile condition, was removed and replaced, with great attention paid to historic accuracy. [Figures 9.10 and 5.2]

The observer's chair at the Detroit Observatory, a device which permits the astronomer to assume the proper position relative to the telescope's eyepiece, was originally placed on a track affixed to the dome room wall just below the cannon balls. This chair rotated with the dome as it was turned, but it limited the observer's maneuverability. At the same time the cannon balls were removed in 1890, this chair was replaced by a free-standing chair of the Burnam-Hough pattern. Then, in 1907, a lighter and less cumbersome observer's chair was purchased, and it is still in use. [Figure 9.11]

Although they were not originally planned for the building,[268] two wings were added, one to house the meridian circle, and the other, for symmetry, to accommodate the astronomer. It is likely that Tappan originally expected to purchase a transit instrument in Europe that was smaller than the one he eventually selected because, upon receipt of the telescope, the west wing had to be redesigned and enlarged to accommodate the meridian circle "on account of the unusual size of the . . . Instrument."[269] There is structural evidence to document that the two wings were not originally planned. The walls between the main building and the wings are three brick courses thick—the same as the north and south exterior walls of the main building; the entryway for the crawl space under the Meridian Room is shaped in a way that suggests that stones from the foundation were removed in order to create the opening; and, perhaps most convincing, decorative metal grates exist in the basement walls between the main building and the crawl space under the wings that appear to be intended for exterior ventilation.

Each wing is 20 feet wide, 18 feet long, and 20 feet high. The west wing contained the books and charts of the Observatory, serving as a library and office for the astronomer. This wing was split into two rooms by an east-west dividing wall just south of center.

Figure 9.11. Observer's chair used at the Detroit Observatory since 1907. The original observer's chair was affixed to the dome, so it moved when the dome was rotated. (Patricia S. Whitesell)

Figure 9.12. Crane prepares to lower a 2-ton replica of the original limestone clock pier through the roof hatch into the Meridian Room at the Detroit Observatory in June 1998. The stone pier is inside a steel cage. (Patricia S. Whitesell)

Although this wall was removed at some point, its shadow can still be seen in the plaster at the walls and ceiling. The two windows in the south half of the wing were plain, lacking the elaborate pediment that is placed over the other windows, which suggests that the north half of the wing served as the more formal library and office, with the south half serving as a computing room and quarters for the astronomer. The east wing contained the meridian circle telescope, two collimating telescopes, and an astronomical clock. Five stone piers were put in place to support the scientific apparatus: two piers, joined together below the floor level, for the meridian circle; two piers, at the south and north ends of the room, on which were mounted the two collimating telescopes; and a pier on the south-facing side on which was mounted the sidereal clock.

The clock that was mounted on this pier was made in Berlin by M. Tiede. [see Figure 4.3] In this location, the Tiede clock was subjected to significant temperature changes, so it was later moved and mounted on the east face of the central refracting telescope pier. The clock pier in the Meridian Room was then removed, and the door to the room was relocated to the north, perhaps to provide a more spacious entryway. The exact date of these changes is not known. Scars were discovered on the central telescope pier that document the exact mounting location of the Tiede clock, and similar scars were found on the south face of the pier where another astronomical clock was located. Also found were electrical wires, snipped just below floor level, that connected the Tiede clock to the chronograph located in the west wing. On the north face of the refracting telescope pier was found a large scar (area of infill) that corresponds with the size of an astronomical clock made by the E. Howard Company that the Observatory obtained around 1893.[270] It was typical to mount clocks on piers in order to limit vibration that would interfere with the functioning of the pendulum, so it is likely that a hollow was created in the north face of the brick pier to accommodate the Howard clock before it and the other clocks were moved to a special clock room in the Observatory's 1908 addition (see Chapter 10). As part of the 1997-98 restoration of the building, the missing clock pier was replicated in limestone, and the 2-ton block of stone was lowered into place through the hatch in the roof of the Meridian Room. [Figure 9.12]

The meridian circle telescope was mounted in the east wing of the building on a pier made of brick below floor level with separate, massive limestone piers above. The telescope hung between the limestone piers on brackets attached to the stone, thereby allowing it to swing from north to south but in no other direction. This achromatic telescope was eight feet in focal length and 6½ inches clear aperture. It pivoted on a horizontal axis to enable the observer to track stars as they crossed the meridian.

The Meridian Room roof could be opened to the sky [Figure 9.13] by raising the two 29-inch wide panels, divided in the middle, that extended the length of the east wing from north to south. These panels were raised and lowered manually by means of a ratchet with pawl and wheel connected to ropes that passed through the roof, then over pulleys attached to the soffit, and then looped over two iron brackets resembling the horns of a longhorn steer that were attached to the exterior of each hatch panel. [see Figures 9.5 and 9.14] Side panels at the north and south walls of the Meridian Room could be dropped down in grooves in the casings to permit observations below the plane of the horizon.

The building's roof was tin, with standing seams on the two wings and batten seams on the central structure. [see Figure 9.7] During the 1997-98 restoration of the building, the asphalt roofing and fiberglass insulation that had been added over the original roof in the early 1970s was removed to reveal the original tin roof. The Meridian Room hatch, long inoperable, was opened and made functional. [Figure 9.14] After photographically documenting the

Figure 9.13. Detroit Observatory circa 1895, one of only two known historical photographs that show the Meridian Room's roof hatches open. Note the windvane (still extant) and anemometer. The original of this image is a cyanotype. (Quinn Evans/Architects)

Figure 9.14. Meridian Room roof hatch was unsealed and reopened in 1997 during the restoration and made functional again, having been sealed shut since the early 1970s. (Lloyd Baldwin)

original roof, it was removed. Under the tin panels were found two different manufacturers' marks: Moorhead & Co., and Griffiths, Wright & Co. Both companies were located in Juniata, Pennsylvania, a suburb of Pittsburgh, though the roofing material would have been purchased through a local vendor. Pittsburgh was a center for the iron industry, with the first complete iron rolling mill in the U.S. established there in 1819.[271] Blooms, which were the bars of steel prepared for rolling, were shipped from Juniata to Pittsburgh, then rolled into flat sheets and coated with lead. The roofing material, known as terneplate, was then distributed around the country by canal and railroad.

The 1997 roof tear-off on the east wing of the Observatory over the Meridian Room confirmed what had been alluded to in the Regents Proceedings: this section of roof had been replaced around 1866 because it leaked.[272] The original roof over the rest of the building remained in amazingly good condition for more than 100 years, but it is not surprising that the Meridian Room roof failed given the complexities involved in making the meridian hatch watertight. The roofing material of the replacement roof, uncovered during the 1997-98 restoration, was galvanized metal instead of tin, which confirms that it had been added at a later date. During the restoration, the name of the installer, C. Fischer, Ann Arbor, MI, was found written in black paint under one of the galvanized panels. [Figure 9.15] The dome's clerestory[273] was also re-roofed, at some point. [see Figure 9.7] The materials for this repair were obtained from the Buhl Sons Company in Detroit,[274] a fact confirmed by an imprint discovered on the roofing material. This roof repair may have taken place when the dome wall was strengthened in 1865.

The Observatory site was gradually developed after the building was constructed, first with the digging of a well and a drain by J. Killmartin for $40.25 (June 1855); then the erection of a fence by D. E. Wines, with fence posts provided by James Black,[275] [see Figure 9.2] to enclose the perimeter of the Observatory grounds at a cost of $193.75 (December 1855); followed by the sodding of the surrounding embankment (October 1857); and the construction of a wood-house (October 1857). Trees were planted all around the campus, with contributions from students, the University, and the faculty.[276] One of the trees planted in front of the Observatory—a horse chestnut—is still extant. [Figure 9.16] It can be seen as a sapling in a circa 1856-58 photograph of the building [see cover or Figure 14.1], which makes it about 140 years old in 1998.

A road to the Observatory was badly needed. Land for a road had been purchased from Julia and Philetus Stark in November 1853 at the time the central parcel of land for the Observatory structure was obtained. In October of 1854, Professors Winchell, Douglass, and Fasquelle were appointed as a Committee to contract for and superintend the construction of a road "from the highway[277] near the Observatory grounds to the Observatory as soon as right of way [was] obtained." This was rescinded in June of 1855 when the City of Ann Arbor agreed to construct a road.[278] Observatory Street continues to provide access to the building today.

Figure 9.15. Roofers discovered during the 1997 restoration the signature of "C. Fischer, Ann Arbor, Mich" painted under a section of the Detroit Observatory's tin roof that had been repaired by Fischer in 1866. The artifact was saved for the Observatory's museum. (Lloyd Baldwin)

Figure 9.16. Detroit Observatory as it appeared prior to its 1997-98 restoration. (Patricia S. Whitesell)

The building's stone foundation walls had leaked from the day the building was built, and water actually flowed into the cellar.[279] Evidence that there was significant flooding in the past was found during the 1997-98 restoration work, especially in the 5-foot deep pit around the central telescope pier, where spalling of the lower eight courses of brick was extensive and mortar joints were completely eroded. The lower courses of brick around the base of the brick archways[280] surrounding the pit had similar water damage. The entire foundation was repointed with "water lime" in the fall of 1865, and the ground around the perimeter of the building was graded so that water would drain away from the structure. At some later date, floor drains were installed to control flooding. As part of the restoration in 1997-98, past damage was repaired and extensive measures were taken to provide proper drainage, yet problems continue.

A bathroom was added to the building, the exact date of which was unknown until the 1997-98 restoration. In the process of expanding the size of the bathroom to permit handicapped access, the signatures of the carpenters who built the wall, Charles A. Hardy and Edward W. Hatch, and the date, January 26, 1906, were revealed. In 1896, Asaph Hall, Jr., the director of the Observatory, proposed to have plumbing work done in order to put a bathroom in the Observatory, provided the Regents would authorize $30 for the carpentry work.[281] Why the walls were not constructed until 1906 is not known. A small window was added in the south wall to permit light and ventilation. The original porcelain sink was still in place at the time of the restoration, though it was necessary to replace it in order to comply with Americans with Disabilities Act requirements.

While the 1997-98 restoration of the Observatory will preserve the building's early history, it will also preserve a later period in the building's life. The basement utility room contains day-glow art work left by "hippies" who occupied the space during the 1970s. These graffiti will be preserved. In 1996, a former University student shared the story that, in the 1970s, a sympathetic astronomy student obtained an unauthorized key to the building and offered it to a man named Joshua, a left-wing counter-culture activist, who needed a place to live. Joshua, who apparently resided in the Observatory's basement for about a year, left behind a graffito that reads "Here Comes the Sun," another that creates the semblance of a closet through the image of a coat hanger painted on the wall, and other day-glow art characteristic of the era. [Figure 9.17] Some years later, the former student encountered Joshua in San Francisco, where Joshua had become a leader of the followers of the Reverend Sun Yun Moon ("Moonies"). Other alumni have reported that the Observatory was used during the 1970s as both a "crash pad" and a place for lively parties. Entry was gained through an unlocked hatch in the roof, and the dome room was a favorite site for regular gatherings. The telescope's wooden base was adorned with day-glow paint, mattresses provided seating on the wooden floor, and the circular wall was decorated with a day-glow mural. Fortunately, little lasting damage was done to the building during this time.

In the 1980s, efforts were undertaken to refurbish the building and to give it an active life. The west wing, which had served in the 1940s and '50s as the Astronomy Library, was adapted for use as classroom and meeting space. The built-in wooden bookcases[282] were removed. [Figure 9.18] The second level was converted for office use by adding modern bookshelves and removing a small ladder that provided access to the roof through a hatch in the ceiling. The building was used as the headquarters for the Collegiate Institute for Values in Science and as a meeting place for the History & Traditions Committee. Repairs were made, including painting of the dome, installation of heat on the second floor, asbestos removal from basement pipes, refinishing of floors, and interior and exterior painting. The Observatory presented a favorable impression to astronomers who assembled in Ann Arbor for the annual meeting of the American Astronomical Society in 1989. Subsequently, the building was used as

Figure 9.17. Graffito "Here Comes the Sun" in day-glow orange and green was painted in the 1960s in the basement of the Detroit Observatory. (Patricia S. Whitesell)

Figure 9.18. Detroit Observatory's Library as it appeared in 1920. (Bentley Historical Library)

a warehouse for historic scientific and medical instruments, which saved them from being discarded, with foresight for future museum exhibits. A team of architecture students conducted a study that documented the condition of the building, which assisted in identifying further work that was needed in order to fully restore and secure the building. These actions helped to stabilize the building until a complete historic restoration could be undertaken.

It was in 1994 that the History & Traditions Committee approached Vice President for Research, Homer A. Neal, to ask whether his office was interested in adopting the building. The proposal written in response to this request outlined a plan for restoration of the Observatory. It was positively received, and after final approval was granted by the President, oversight of the building was transferred to the Office of the Vice President for Research. A year was spent clearing the building of the artifacts held there in storage, cleaning, and providing temporary furnishings so that the building could be used for activities in support of fundraising for the building's restoration. Research on the building's history ensued, the fruits of which have resulted in this book. Vice President Neal appointed an Observatory Advisory Group whose first task was to conduct a national search for a historic preservation architect. The search led back to Ann Arbor, with the selection of the firm Quinn Evans/Architects, whose local knowledge and national reputation for work on historic sites made it a good match. A preservation study was completed by the architects, assisted by research provided by the Advisory Group, and extensive planning led to the preparation of detailed architectural plans. A general contractor was selected by competitive bid, with J. C. Beal Construction gaining the contract.

The construction phase of the project, which was estimated to take 9 months beginning in June 1997, actually took about 15 months. The objective was to stabilize and secure the building, and to restore it with historical accuracy to its early period. [Figure 9.19] For stability, foundation repairs were made at the southwest corner, waterproofing was performed along the south foundation wall, and some additional shoring posts were installed in the basement. Extensive tuckpointing was needed at the basement and foundation. The electrical wiring, which was the old knob-and-tube style, was deactivated but left in situ for historical interest, as were old telegraph and other wires. New wiring was added to bring the building up to modern codes. This work was extensive and required intrusion behind plaster ceilings, under wooden floor boards, and in some cases, channeled into brick walls in order to hide the modern wires. In the course of this work, evidence was found of habitation long ago: an enormous, inactive paper wasp nest was discovered in the second floor ceiling rafters. A honeybee hive, which was about six feet long and two feet wide, was removed from the cornice at the southeast corner of the building. Because honeybees are endangered in Michigan due to mites, and these bees were mite-free, they were relocated by a beekeeper using pheromone bait to lure them into a portable hive. The old hive was then removed from the cornice. [Figure 9.20]

To protect the building from intrusion, a security system was installed. Handicapped access was provided with historic sensitivity to permit access to the building and its restroom facility. The concrete steps leading up to the

Figure 9.19. Workers install a sign at the Detroit Observatory to announce the 1997-98 restoration and interpret the history of the building for passersby. (Lloyd Baldwin)

Figure 9.20. Honeybee hive removed from the cornice of the Detroit Observatory in 1995. (Patricia S. Whitesell)

building from East Ann Street were replicated from historic photographs, providing a breath-taking entry point to the building's site. [see Figure 10.7] Paint was stripped from the dome in order to locate leaks, then the dome was repaired and painted. The metal roof was replaced using lead-coated copper, replicating its original appearance, and the dome shutter and meridian hatch were made functional and water-tight. [Figure 9.21] The Meridian Room door, which had been relocated ten feet to the north, was moved back to its original location, and a window in the director's office/library that had been removed when the director's residence was added in 1868 was put back in place. Paint analysis was performed throughout the building to determine the original colors used in the 1850s, and these historic colors were used for a complete interior and exterior paint job. Exterior window shutters and the balustrade over the front porch, both long missing, were replicated from historical photographs. Exterior lighting, which can be turned off while the telescopes are in use, was added to highlight the building's facade at night.

After an extensive search, museum consultant and historic telescope restoration specialist Christopher Ray was engaged to restore the Fitz refracting telescope. A Telescopes Subcommittee, which was formed to provide technical expertise to the Advisory Group, recommended after lengthy discussion that the Fitz refractor be restored to its 1907 condition instead of attempting to replicate its original wooden tube. This decision was reached because insufficient knowledge exists regarding wooden tube technology to be sure the telescope would function properly if a replica of the wooden tube were made to replace the current steel tube. Restoration to its 1907 appearance was successful, and its original brass components sparkle once again against a glossy black, steel tube. [see Figure 5.7 in centerfold] A scale model which replicates the telescope's original appearance will interpret the instrument's original

configuration. The telescope's unique Fitz lens will be professionally documented and cleaned and then placed back in the telescope to enable
astronomers, students, and guests to once again view the stars and planets.
Research on the Pistor & Martins meridian circle telescope revealed that it
possesses such a degree of historical significance that its restoration will
be postponed until the research has been completed, in order to ensure that
proper decisions are made. Both telescopes were protected during the construction phase of the restoration by building around them plywood crates
which could withstand falling plaster and other threats.

An extensive search of the campus was successful in locating numerous
items of historical, scientific interest to the Observatory. Scientific instruments
and other artifacts related to the Observatory were located in storerooms,
faculty offices, classrooms, residences, museum basements, and even in the
Observatory: a thermometer, tucked long ago under some stairs to protect it,
was found during the restoration work. The artifacts will be used to create
museum displays that interpret the building's history and its scientific
activities. Bookcases that encircle the central telescope pier will be utilized as
display cases for scientific instruments and other artifacts, with the telescopes
being the primary museum attractions.

Figure 9.21. Workers make repairs to the Detroit Observatory's dome during the 1997-98 restoration. Note the newly completed roof replication on the west wing. (Lloyd Baldwin)

Figure 10.1. Stereoscopic photograph of the Detroit Observatory showing the 1868 director's residence addition. James C. Watson (wearing top hat) is seen with a telescope on a tripod; his wife, Annette, is standing at the porch. Sapling in front of the Observatory has grown (see cover and Figure 14.1), and is today a magnificent Horse Chestnut (see Figures 9.16 and 14.2). (Collection of Leonard A. Walle)

Chapter 10

Additions and Relocation Attempts

Numerous attempts were undertaken over the years to relocate the Observatory structure—so many attempts, in fact, that it is remarkable the building still stands in its original location. Efforts to move or demolish the building began as early as 1862, only eight years after its original construction, and continued as late as 1976.

Although the Observatory was a source of considerable pride for Tappan, neither Tappan nor Brünnow were ever satisfied with the building's location. After leaving the University in 1863, Tappan wrote of the Observatory:

> The conception of it was my own. I proposed the erection of it. Aided faithfully by dear friends in Detroit, I obtained subscriptions for it. Traveling in Europe, I visited various astronomers, and under their advice, and with the special aid of Encke and Brünnow, procured the instrument which constitutes its great value. I procured all the instruments except the Chronograph, brought by Brünnow from Albany. I disbursed the money. I am responsible for everything but its location. This was determined while I was in Europe, without my advice or concurrence. It has proved an inconvenient location, and has caused much fatigue to the astronomer. But the site is beautiful; and the prejudice which selected one of our highest hills may be pardoned . . .[283]

It was in March of 1862 that the Committee on the Observatory was charged by the Regents to investigate the expediency of relocating the Observatory to the University grounds. Brünnow was added to this Committee. Brünnow had always been dissatisfied with the building, despite its design by Professor Bull who was knowledgeable about astronomy. In 1866 after leaving Ann Arbor, Brünnow reflected in a letter to Andrew D. White:

> I hope you do not consider it arrogant of me if I render to give you a little advice in regard to the [Cornell] observatory, in order to prevent you to fall into the same mistake, which has been made so far in all cases when an observatory was built in your country, namely this, to erect the building before an astronomer has been appointed. I know from experience how much I suffered from this very mistake having been committed at Ann Arbor. If I had been there before the building was erected, I could have made for better arrangements and saved at

the same time a good deal of money. I therefore advise you by all means to let this first thing be the appointment of an astronomer, and let him make the plans for the observatory.[284]

Following several months of study, the Committee on the Observatory reported that the cost to erect a similar building would be $4,200, and that about $2,500 could be gained from the sale of the current premises. Brünnow presented plans of his own design for a greatly improved building, but without any estimate of the additional cost.[285] The Regents approved the plan, with the understanding that a condition of the contract be the acceptance by the contractor of the present building and site, after all salvageable parts were incorporated into the new structure, as part of the payment. This proposal was never acted upon because Tappan was removed from office in 1863, with Brünnow's resignation following shortly after. Had Brünnow stayed at the University, it is likely that the original Detroit Observatory structure would not have survived.

Watson, who took over upon Brünnow's departure, again proposed the relocation of the Observatory in August of 1865. The proposed site for the new Observatory was near the southeast corner of the 40-acre University grounds, or roughly where West Hall (formerly the West Engineering building) currently stands. The rationale for relocation of the building was based on factors such as the disturbance caused by vibration from passing trains; the improper construction of the original buildings; the lack of fire protection incorporated into the building for the protection of the instruments; the lack of provisions for making and recording magnetical and meteorological observations; impaired visibility; dust and smoke in the vicinity of the original structure; and interruption by visitors.

The Regents approved Watson's plan, contingent upon funding for the project being raised from the City of Ann Arbor, the citizens of Detroit, and the sale of the present lands. The City agreed to provide $10,000 toward the project if an equal amount could be raised, but it was subsequently decided that the Observatory would, instead, be repaired and an addition added to accommodate the director.[286] The reason for this change of plans is not known. It is possible that the citizens of Detroit were unwilling to contribute toward the project because they were angry about the dismissal of President Tappan that took place in 1863 [see Chapter 13]. The City of Ann Arbor provided $2,500 toward the repairs and $500 toward improvement of the streets around the Observatory, and Watson secured $3,000 in subscriptions for the required match. The contract was awarded to Messrs. Morrick and Bliss, and work proceeded on construction of a dwelling for the director, with Professor Douglass superintending the construction, at a total cost of $5,075.[287]

It was when Brünnow left the University in June of 1863 that Watson asked the Regents to provide a residence for himself and his family. Watson had built a house on South University Ave. when he married in 1860. He had spent a great deal of time in this endeavor, being involved in every detail, including placement of orders for building materials. His participation in the process, no doubt, saved him money. Nonetheless, he found himself in serious debt. One way of alleviating his financial problems was to have his housing provided by the University, thereby permitting him to sell his home to pay off debts, and live rent free. Although it appears that Watson sold his home in 1862 (see Chapter 11), it wasn't until September 1865 that Watson obtained the Regents' approval to build a residence at the Observatory.

In contrast to Watson's approach to obtaining accommodations, when Brunnow returned to Ann Arbor from the Dudley Observatory in 1860 [see Chapter 12], he requested and received approval from the Regents to purchase from the University about 2½ acres of land on the west side of the Observatory on which to build a residence for himself.[288] It is not known why this plan was never followed through. Apparently, Brünnow had dropped the notion of relocating the Observatory to central campus, though it is suspected that a hazing Brünnow and Tappan received through the press contributed to the change of plans. However, it is noted that Brünnow's desire to build a residence in 1860 is an indication of his inclination, at that time, to remain at Michigan.

The director's residence, completed in 1868, was added to the west wing, with access to the original building provided through a door where the southwest window opening had been located. [Figures 10.1 and 10.2] Watson made

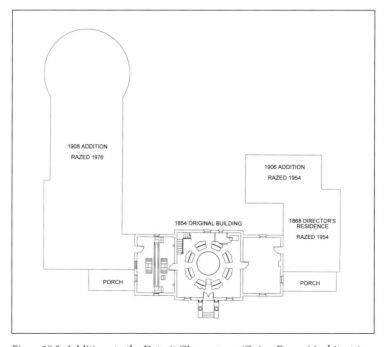

Figure 10.2. Additions to the Detroit Observatory. (Quinn Evans / Architects)

Figure 10.3. Postcard showing the Students' Observatory, which was located to the south of the Detroit Observatory. (Quinn Evans / Architects)

a few improvements to the original building in 1869, adding bookcases and a case for the chronograph at a cost of $54.25.[289] A furnace was requested for the residence in June of 1882, and completed the following year, with the work done by C. Eberbach, and the coal house was converted the following year to function as a barn.[290] The residence was considerably enlarged and updated in 1905, when William J. Hussey was appointed as director. [Figures 10.2 and 9.13] The Observatory residence was continuously occupied by the director until 1942, when it was used as a residence for women. Then in 1946, it was renovated and divided into three apartments for the director and members of the astronomy faculty, and was eventually occupied by faculty from other departments. When Couzens Hall was expanded in 1954, the Observatory residence was razed to make way.[291]

A Students' Observatory was erected in 1878, during Watson's administration, to alleviate competing demand for use of the telescopes. [Figure 10.3] Watson had started the project in response to complaints that the Observatory's telescopes were seldom available for student use. To fund the project, Watson had obtained the Regents' approval to accept an offer made by the United States Government to erect a small observatory in order to observe the transit of Mercury, which would take place in May 1879. The cost to the University for this undertaking was less than $200.[292] The building, constructed just south of the main building, was equipped by the Government with a refracting telescope and a transit telescope, and a small pier was constructed about 60 feet south of the Students' Observatory for photographic operations related to the transit of Mercury. It held a reticle plate and photographic plate holder for a heliostat and camera lens. [Figure 10.4] The Students' Observatory building was used again in 1882 to observe the rare transit of Venus,[293] although cloudy skies made observations only partially successful.[294] Several other outbuildings were constructed on the Observatory grounds to house an instrument shop and other adjunct functions.

The Government reclaimed its instruments after the transit of Mercury observations took place. Not long after Mark W. Harrington took over as director of the Observatory in 1879, he appealed to the Regents for funds and was granted $3,050 for the purchase of comparable instruments: $1,800 for a 6-inch equatorial telescope, $1,000 for a 3-inch transit telescope, and $250 for a chronometer.[295] Harrington selected a 6-inch refractor by Alvan Clark & Sons, a 3-inch transit telescope by Fauth & Co. (with Clark optics), and a chronometer by Elgin.[296] [Figures 10.5 and 10.6] These new instruments were used exclusively for instruction, which freed the large telescopes for research use.

Competing demand for time on the large telescopes was thus relieved, but the desire to relocate the Observatory persisted. In 1903, Cleveland Abbe, an 1860 Michigan graduate who was trained by Brünnow and went on to become

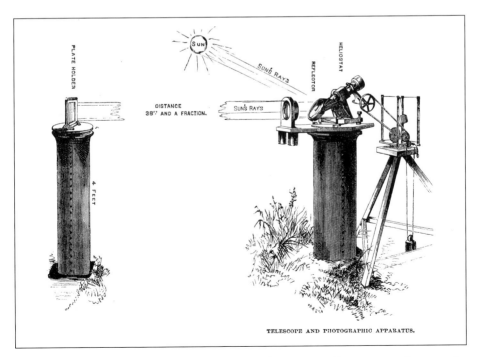

Figure 10.4. A camera similar to this one was used at the Students' Observatory to document the transit of Mercury in 1879. (*Harper's New Monthly Magazine,* 1874)

Figure 10.5. Refracting telescope (6-inch) by Alvan Clark & Sons was mounted in the Students' Observatory. (Bentley Historical Library)

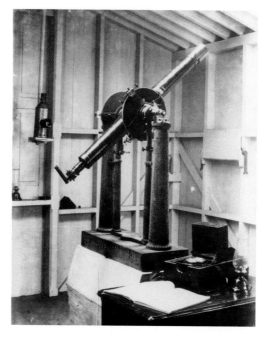

Figure 10.6. Transit telescope (3-inch) by Fauth & Co. mounted on a concrete pier in the Students' Observatory, and an Elgin chronometer (both still extant). Artificial horizon (black object on wall shelf to left of telescope) can also be seen in Figure 2.1. (Bentley Historical Library)

an accomplished astronomer and scientist, prepared an article for the *Michigan Alumnus* magazine that encouraged the University to relocate the Observatory:

> . . . it is evident that the Observatory must be removed to a better location [so] the Director can do the best work that his instruments are capable of. During the last fifty years very many observatories have been thus forced to remove or rebuild. Those at Washington and Cincinnati are good examples. At the present moment the magnetic observatory at Washington is rendered useless by the system of electric trolleys. A university must always hold itself in readiness to keep its educational and scientific apparatus and their environment up to the highest standard that the progress of science requires. Michigan and her alumni should not allow her observatory to fossilize.[297]

But, the Observatory held firm in its location. Instead of moving it, a formidable addition was built onto the east wing in 1908. [Figures 10.7 and 10.2] This new facility included classrooms, offices, a clock room, and a 37½-inch reflecting telescope intended for spectrographic work. It was one of the largest telescopes in the world at the time, made entirely in the Observatory Shop, except for its mirror. [Figure 10.8]

Figure 10.7. Detroit Observatory circa 1927 with director's residence to the right and 1908 addition to the left. Smoke from Power Plant the University built circa 1912 (see smoke stack) interfered with telescope observations. (Bentley Historical Library)

Figure 10.8. Reflecting telescope (37.5-inch) was constructed entirely in the Observatory Shop and installed in the dome of the 1908 addition. (Bentley Historical Library)

In 1908, the first encroachment on the Observatory's land took place when the University proposed to grade the land just west of the building for use as a women's athletic field. An appeal to President Angell was successful in halting the plan. To protect the land to the east, Robert P. Lamont and his wife gave the University in 1911 approximately 26 acres of land adjacent to the original site, with the objective being to head off encroachment by buildings, the intrusion of lights and smoke, and other hindrances to astronomical observations. But, trouble developed to the south when, in 1912, a plan was developed to drain the Cat-Hole and build on that site a new power plant for the University. Despite protestations, placement of the power plant at this location was considered too important. A new site for the Observatory would need to be found because the smoke generated by the power plant's chimney interfered with observations through the telescope. [see Figure 10.7]

A site called Huddy Hill was considered for the Observatory, and considered again in 1915 when a new hospital was planned to the north. No action was taken, but another encroachment took place in 1922 when land to the west of the Observatory was selected for a new dormitory for nurses, called Couzens Hall. Not only was the land consumed, but the Students' Observatory, which had been moved to the west to make way for the 1908 addition, had to be dismantled to make way for the dormitory. The fifth floor of Angell Hall was designated for use by the Astronomy Department, and two domes were added to the roof of that building for use by students. Two 20-foot domes by J. W. Fecker were constructed, and installation of a 10-inch Warner and Swasey refracting telescope (1926) and a 15-inch reflecting telescope (1929) took place.

The likelihood of continued encroachment on the Observatory site was certain, and further appeal was made to relocate the Observatory. This time, a site west of town on Liberty Road was considered, but action was postponed. In 1924, when a site was needed for the new Simpson Memorial Institute for Medical Research, Robert Lamont granted approval to use a portion of the land to the east of the Observatory. The Lamont land became the subject of dispute in the 1960s when the Lamont Trust was challenged by the Astronomy Department, but this was apparently dropped when a new structure, the Dennison Building, was constructed in 1963 to house the Physics and Astronomy departments. The Lamonts provided further contributions to make possible Dr. Hussey's dream: the Lamont-Hussey Observatory was erected in Bloemfontein, South Africa, completed in 1928. The new 27-inch refracting telescope was set up in July 1925 on the lawn of the Observatory, next to the 1908 addition, to test it before it was shipped to South Africa. [Figure 10.9] Tragically, Hussey did not live to realize his dream: while en route to South Africa, he died in London of pleurisy. Richard A. Rossiter [Figure 10.10], who was in London with Hussey when he died, took over the operation of the Lamont-Hussey Observatory.

Figure 10.9. Lamont-Hussey telescope (27-inch) was set up on the lawn of the 1908 addition for testing before shipment to Bloemfontein, South Africa in 1926. (Bentley Historical Library)

Figure 10.10. Richard Rossiter (right) and colleague using the Fitz refracting telescope at the Detroit Observatory circa 1922. (Bentley Historical Library)

The search for a new site for the Observatory continued until a parcel in the hills near Portage Lake, fourteen miles west of Ann Arbor and adjacent to the University's forest preserve, was selected, but no action took place. By 1928, Dr. Curtiss, the Observatory's new director, must have been deeply frustrated. His subsequent appeal to the Regents resulted in the purchase of the site near Portage Lake, a 200-acre parcel of land at Peach Mountain, and a new observatory was placed as a top priority in the University's appeal to the State Legislature for funding the following year. Pronouncement of the availability of the land occupied by the Detroit Observatory as a potential building site was used to make the deal more attractive.

The financial depression of the 1930s postponed the land purchase until 1945. Eventually, equipment from the McMath-Hulbert Observatory that was established in 1930 at Lake Angelus near Pontiac, Michigan was moved to Peach Mountain, with the two new structures at this site inclusively named the Portage Lake Observatory. When the location of these observatories eventually became unproductive, the University found clear skies by joining a consortium with Dartmouth College and the Massachusetts Institute of Technology (MIT) at Kitt Peak, Arizona, and also with the Cerro Tololo Interamerican Observatory near Santiago, Chile. A new project called Magellan, which is a consortium with the Carnegie Institution of Washington, Harvard University, MIT, and the University of Arizona, is currently under construction at Las Campanas, Chile.

The traffic area in front of the University Hospital (Old Main), located just across the street from the Observatory, became so congested in the 1960s that a plan was developed in 1966 to move the Observatory's original building 20 feet to the south. This would have permitted the slope on which the building is constructed to be leveled, so that two lanes of traffic could cut across the northeast corner of the property on the diagonal.[298] Fortunately, this plan was never carried through. Instead, the Old Main Hospital was razed in 1989 and a new hospital was constructed at an adjacent site to the northeast.

In the 1970s, plans were developed to raze the Detroit Observatory and its 1908 addition, prompted largely by the deteriorating condition of the 1908 addition, and the disuse of the facility. Construction of the Dennison Building in 1963 to house the Physics and Astronomy Departments, and the use of distant observatories, rendered the old facilities and telescopes obsolete. The proposal provoked considerable resistance from within the University and local communities. A group led by local preservationist John Hathaway and Astronomy Professor Hazel "Doc" Losh was successful in dissuading the University from razing the original 1854 building. However, the 1908 addition had fallen into such a state of disrepair that it was condemned as unsafe. Only the original building was considered to have historical value. The 1908 addition was thought to detract from the observatory complex by obstructing the view of the original building from the east. With general agreement, the 1908 addition was demolished in 1976. [Figure 10.11] The massive 37½-inch reflecting telescope was disassembled, but several of its parts, including the mirror and driving clock, are in storage at the University. The mount was given to the Lake Erie Astronomical Project.[299] After the demolition, the cornerstone of the 1908 addition was tucked away in the basement of the Detroit Observatory. In 1998, to commemorate the 1908 addition, the building's cornerstone was built into the south end of a stone wall, which is located where the 1908 building once stood.

The Detroit Observatory building was spared, and it persists today in its original location, having held tight to its original footings and instruments, despite so many close encounters with relocation over the years.

Figure 10.11. 1908 addition was demolished in 1976. (Bentley Historical Library)

Figure 11.1. James Craig Watson circa 1867. (Bentley Historical Library)

Chapter 11

James Craig Watson

James Craig Watson was Franz Brünnow's most promising student at the Detroit Observatory, and he eventually became the Observatory's second director. Watson was born near the village of Fingal, county Elgin in Ontario, Canada on January 28, 1838. [Figure 11.1] His father was a farmer and carpenter, and he also taught school, passing along to local children and his own three sons and a daughter the knowledge he had gained from his father's interest in books. Watson's grandfather had also been a farmer in Upper Canada, with an interest in books and an excellent library, and he had died a wealthy man.[300]

It was in 1850 that Watson's father fell on hard times and was forced to abandon his home. The family traveled to Detroit where a stranger suggested Ann Arbor as a place the family could reach by train. Watson's mother decided to follow the advice, believing that Ann Arbor held an opportunity for her children to become educated. Watson's father found work at a small factory, and James took on various menial jobs, learning quickly by observing. Soon, he displaced a less competent factory engineer, having become a skilled machinist. Every chance he had, Watson studied Latin, Greek, and other subjects, and he excelled at mathematics. When the factory closed, he was reduced to selling apples and books at the railway station, which he found so humiliating that he soon disappeared to Detroit, and was about to sail off with a Great Lakes trader when he was convinced to return home.

At the young age of fifteen, Watson entered the University, graduating with honors in 1857. Watson, who was considered to be a child prodigy in mathematics, was Dr. Brünnow's first student in astronomy, receiving his bachelor's degree in 1858 and his master's in 1859—the first year the master's degree was awarded at Michigan. At times, Watson was the only student in Brünnow's class. Both theoretical and practical astronomy interested Watson, but his mechanical talents lured him into telescope making and lens grinding and polishing, guided by his translation from the German of Prechtl's *Praktische Dioptrik,* and by his correspondence with American telescope maker, Henry Fitz, Jr. At age nineteen, Watson finished work on a 4-inch telescope, complete with mounting stand. As his skills in telescope making improved, Watson

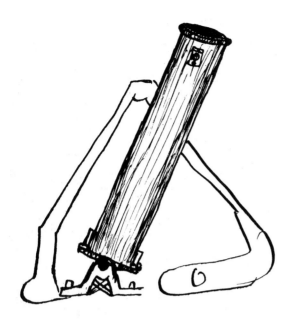

Figure 11.2. Reflecting telescope designed by James C. Watson as sketched in his 1855 jotting book. (Bentley Historical Library)

developed a plan to construct a seven-foot Newtonian reflecting telescope which he intended to build during his summer break. He prepared a written description of this plan in his jotting book which included a sketch of the telescope. [Figure 11.2]

Following his graduation, Brünnow appointed Watson in 1858 as Assistant Astronomer, a position he held while completing his graduate study. Watson went on to make a name for himself, and the Detroit Observatory through his discovery of twenty-two asteroids (minor planets), one of which, Helena, he discovered on August 15, 1868 and named after his wife, Annette Helena Waite Watson. In a congratulatory letter on the discovery of Helena, Benjamin Apthorp Gould remarked "it is pleasant to know that Mrs. Watson is not vexed at having her name taken and does not regard it as taken in vain. It is a fortunate thing that nowadays some folks can go to heaven without abandoning us on earth."[301] Another minor planet, Juewa, which Watson discovered on October 10, 1874 in Pekin (now Beijing), China during a transit of Venus expedition, was named by Watson's Chinese hosts, at Watson's urging.

It was during the year 1868 that Watson discovered six asteroids—a remarkable achievement that won him the Lalande gold medal from the French Academy, awarded in 1870. All of Watson's asteroids except Juewa were discovered at the Detroit Observatory between 1863 and 1877. When Watson discovered his first asteroid in 1863, which he named Eurynome, it was the 79th ever discovered. The very first asteroid identified was in 1801, by the Italian astronomer Giuseppe Piazzi of Palermo; he named it Ceres after the goddess of Sicily where he made the discovery.[302] It wasn't until 1854 that an American was successful in the competition, with Fergusson's discovery of Euphrosyne (number 31) in Washington, D.C. He discovered another, Virginia (50), in 1857, and then Searle of the Dudley Observatory discovered Pandora (55) in 1858. Fergusson met with success again in 1860 with Echo (60), followed by Tuttle's Maia (66) and Clytie (73) discovered at Harvard in 1861 and 1862.

In May of 1861, C. H. F. Peters of the Hamilton College Observatory had entered the contest as a serious competitor in the asteroid hunt, or as Watson called it, "bagging asteroids."[303] Peters discovered three asteroids in rapid succession: Feronia (71) in 1861, and Eurydice (75) and Frigga (77) in 1862. Watson rose to the challenge with his first discovery, Eurynome (79), in 1863. In 1867, Minerva (93) and Aurora (94) were discovered within a few days of each other. Then, in 1868, Watson's discoveries came like a streak of lightning: Hecate (100), Helena (101), Hera (103), Clymene (105), Artemis (106), and Dione (107). Peters had upset the string by adding Miriam (102), no doubt raising Watson's ire because the competition between the two had become serious. The fact that Brünnow had hosted Peters in Ann Arbor in the summer of 1861, while Peters was working on the *Nautical Almanac,* must have made matters worse.[304]

More than once, Watson was challenged regarding the right to astronomical discoveries. An example involving Peters took place in 1865 with the pronouncement by Watson of his discovery of his second asteroid, which he named Sappho. After some debate, he had to accede the claim to Peters, who had observed the planet (which he named Io) before Watson.[305] Between 1871 and 1877, Watson discovered 13 more asteroids, bringing his lifetime total to 22 before he retired from hunting asteroids.[306]

His great success at the discovery of asteroids was due to Watson's remarkable memory and command of detail. He laboriously prepared detailed charts that recorded the position of stars located near the ecliptic,[307] thus helping him to recognize objects that had not previously been identified. After Watson's death, tables on Watson's asteroids were computed and published according to the wishes he detailed in his will, and some asteroids proved to be so elusive that it took many years to locate them all.

Watson published ample numbers of papers and reports (see Appendix 2) on astronomy. The most famous was his textbook, *Theoretical Astronomy,* which was published in 1868. This text contained little original material, yet it was considered exceptional in its thorough coverage of the subject, drawing from every possible source. Its shortcoming was a lack of proper attribution, a deficiency that would recur in other works by Watson. Watson's personal, autographed copy of *Theoretical Astronomy* is held in the Detroit Observatory's museum. The book was given to John Schaeberle, Watson's capable assistant (who, among other achievements, discovered two comets), by Mrs. Watson after Watson's death, then passed to Schaeberle's nephew, then returned to the Observatory library during Hussey's directorship, and recently returned to the Observatory by the University of Michigan Libraries.

Professor Peirce of Harvard, a friend of Watson who was also Superintendent of the United States Coast Survey, funded Watson in 1869 to perform research on the lunar tables. The existing tables in the *American Ephemeris and Nautical Almanac* needed improvement for use in practical navigation, and Watson supervised a team of Michigan alumni to perform the

Figure 11.3. Stereoscopic photograph of 1869 solar eclipse expedition to Burlington, Iowa. Left to right: E. C. Pickering (became director of Harvard's observatory); Donald McIntyre, University of Michigan Treasurer and former Regent; James C. Watson, director of Detroit Observatory; William Johnston, Van Vleck's assistant; John M. Van Vleck of Wesleyan University; and George B. Merriman, Asst. Prof. of Mathematics at Michigan. (Collection of Leonard A. Walle)

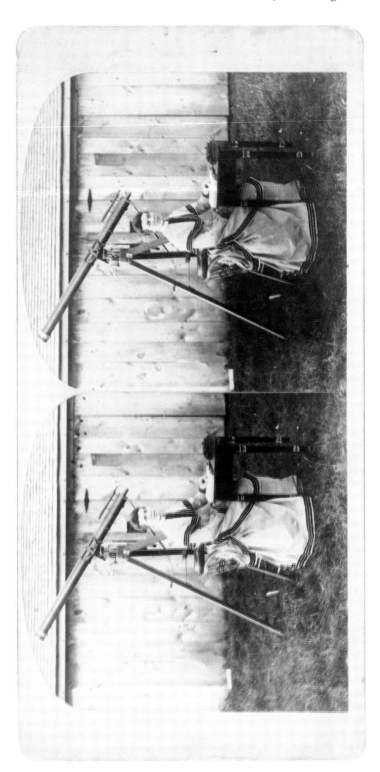

Figure 11.4. Annette Waite Watson, who capably assisted her astronomer husband, accompanied him on this 1869 solar eclipse expedition to Burlington, Iowa. (Collection of Leonard A. Walle)

complicated computations, an enterprise which continued for five years. While the outcome was satisfactory, the tables were never published and were lost.[308]

Practical topics not related to astronomy were also subjects Watson pursued, publishing in 1878 his *Tables for the Calculation of Simple and Compound Interest and Discount*. In 1870, he went into the book and stationery business in Ann Arbor with B. J. Conrad. Then, in 1873, he purchased Dr. A. W. Chase's lucrative Ann Arbor Printing and Publishing Company[309] for $12,500, in partnership with local businessmen Henry Krause, Zina P. King, and Sedgwick and Henry Dean. This company published the internationally famous and enormously popular recipe book *Dr. Chase's Recipes; or Information for Everybody*, in which appeared the recipe for the stucco used for the Observatory. Examples of other recipes included in the book range from toad ointment to various types of glue—a recipe for virtually every need. When Watson departed for Pekin, China for a transit of Venus expedition, Dr. Chase sent him a prescription to treat cholera and diarrhea.[310] In an early edition of the recipe book (1864), Watson provides a testimonial to the book's merits: ". . . it is a work which should find its way into every family in the land."[311] Included in the book are copyrighted tables of interest calculations that, although there is no attribution, are surely the work of Watson. Most likely, Watson sold the rights to the tables to Dr. Chase, and upon Chase's retirement, when Watson and others took over the business, Watson was then free to separately publish his expanded *Tables for the Calculation of Simple and Compound Interest and Discount* (1878).

Watson participated in numerous important astronomical expeditions. In the company of several other prominent astronomers, he participated in an expedition to Mount Pleasant, Iowa in 1869 to observe a total eclipse of the sun. [Figure 11.3] Watson was accompanied by George B. Merriman, Assistant Professor at the University; Donald McIntyre, Treasurer of the University, former Regent, and amateur astronomer; Professor John M. Van Vleck of Wesleyan University, a civil engineer, son of a former University of Michigan Regent, and amateur astronomer;[312] Van Vleck's assistant, William Johnston; Professor E. C. Pickering[313] of the Massachusetts Institute of Technology; Professor Henry Morton[314] of the University of Pennsylvania; and Watson's wife, Annette, who capably assisted her husband in his scientific endeavors. [Figure 11.4[315]]

Merriman observed the eclipse using the Detroit Observatory's Fitz comet seeker, McIntyre used the finder from Michigan's 12⅝-inch Fitz, and Watson used a Clark 3½-inch portable equatorial telescope that Van Vleck had brought along. The telescopes were set up on the grounds of the Henry County Agricultural Society.[316] Professor Morton distinguished himself on this day by conclusively proving that a bright line that appeared on photographs of the sun during the partial phase of an eclipse was a photographic rather than optical phenomenon.[317] [see Figure 1.4] Professor Pickering, using a portrait camera, had the honor of taking the first successful photograph on that day of the corona of the sun.[318] Two graduates of the Michigan Engineering

Department, B. C. Alexander and Frank Krause, were also present that day, one assisting by calling out the beats of the chronometer, and the other recording the instants of the exposure of photographic plates.[319]

In July 1878, Watson participated in another solar eclipse expedition to Separation, Wyoming, borrowing a 4-inch Clark refracting telescope from the Michigan State Normal School (now Eastern Michigan University). Many famous astronomers and scientists traveled to the western frontier for this eclipse, including Dr. Henry Draper of New York, who was accompanied by Thomas A. Edison, the famous inventor. It was largely a vacation adventure for Edison, who was 31 years old at the time, though he also had a professional interest. He intended to test one of his inventions—a tasimeter—which was designed to measure infrared radiation. In this case, the tasimeter would measure the temperature of the solar corona during the eclipse.[320] During the eclipse, astronomers hoped to learn more about the sun's corona to determine its composition and characteristics. Annette Watson participated in the eclipse party's scientific work by making a sketch of the corona just as the eclipse reached totality.[321] [Figure 11.5] She took a great interest in astronomy and was

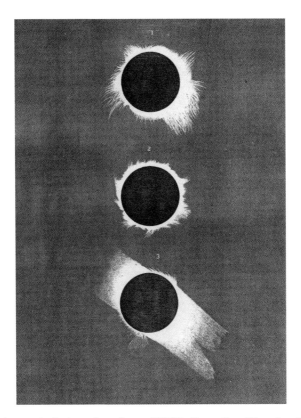

Figure 11.5. Sketches made during solar eclipse of 1878 in Separation, Wyoming. Top sketch (#1) is by Annette Waite Watson. (U.S. Naval Observatory)

a regular participant in and contributor to Watson's scientific work. Several other people made sketches, done according to Watson's instructions, in an effort to document the configuration of the corona around the sun. The sketches were dissimilar, indicating that human observations of solar phenomena are inherently subjective, and therefore unreliable.

It was in Wyoming, while Watson was sweeping the sky during the eclipse looking for objects visible near the sun, that he became convinced he had discovered two intra-mercurial planets, one of which he hoped was the elusive planet Vulcan. The existence of a planet positioned between Mercury and the sun had been postulated by the French astronomer LeVerrier, and in 1859, a French physician, Lescarbault, thought he had seen the planet in its transit across the sun. Many astronomers sought to document the existence of Vulcan, including Watson who had corresponded with LeVerrier to obtain the predicted times of transit. Conditions during an eclipse presented the ideal opportunity to look for Vulcan. In Wyoming, the telescopes were in place. Watson knew he had to make his observations quickly:

> . . . I went back to the sun, moved the telescope nearly one field south and swept out again toward the west. In this sweep I came across a bright star, also ruddy in appearance, which arrested my attention, and for fear that the sun might reappear before I could make an examination of its surroundings, I determined to make a record of its place upon my circles. This I next proceeded to do, and just as I had completed the record the sun reappeared.[322]

Watson announced his discovery by telegraph and was inundated with mail by the time he returned to Ann Arbor. Another astronomer, Lewis Swift, who had been stationed at Denver during the eclipse, reported seeing Vulcan, but the positions reported by the two astronomers conflicted. Most astronomers believed the two had mistaken stars for the elusive planet. Watson's claim to the discovery was controversial, to say the least, and he was never able to confirm his sighting. Watson devoted significant energy to proving his claim, up until his death in 1880. The matter was so controversial, and the details so voluminous, that an entire book called *In Search of Planet Vulcan* was written to chronicle the controversy.[323] The controversy was finally put to rest in 1929, with the conclusion that Vulcan did not exist.

Pekin, China was Watson's destination in 1874-75 to observe the rare transit of Venus, after which he and his wife, Annette, traveled around the world from west to east. While in Egypt, at the request of the Khedive, he surveyed the base line of that country, measured the pyramids, and taught mathematics to an officer of the Royal Guards. His was recognized with decoration as a Knight Commander of the Imperial Order of the Medjudieh of Turkey and Egypt, and with the loan of a houseboat for a trip up the Nile.[324]

Watson was regarded as brilliant. He was facile with numbers, and he used this skill to great advantage in the study of astronomy. Under the able tutelage and mentoring of Franz Brünnow, Watson was staged for an outstanding career. When questioned as to why Brünnow would teach a class to the benefit of only one student, Brünnow replied, "That class consists of Watson."[325] Yet, Watson's reputation was repeatedly tarnished by his obsession with the pursuit of financial gain, and by a general lack of tact. Watson's fault was attributable to his desire to achieve without expending effort, and to a lack of patience for his advancement to happen in due course.

For Watson, lucrative endeavors took priority over the pursuit of science. In the early 1860s, he became involved in the reduction of the Washington Zones for the U.S. Coast Survey, which involved the exacting determination of star positions using time-consuming, complicated mathematical calculations. Although he could perform this work with greater rapidity than others, it was work that Watson undertook for the compensation. Many thought Watson was wasting his talents—talents which were truly extraordinary. He was continually in financial trouble, and his reputation in this regard was widely known. When Watson married Annette Waite of Dexter in 1860, he built himself a house in Ann Arbor on four lots he purchased on South University Ave., at a total cost of about $5,000.[326] After assembling all the resources he could, and borrowing $2,000, he was still severely short. In October of that year, he wrote to Professor Andrew D. White requesting a mortgage in the amount of $2,500 at 7 percent interest, even suggesting that White's friends might be asked to contribute if White could not cover the amount himself. The house was to be the collateral. In January 1861, Watson again wrote to White asking for a loan so that he could pay taxes due that very day.[327]

Watson wasn't the only University faculty member requesting financial assistance from Andrew D. White, who was made wealthy through an inheritance from his father, a banker in Syracuse.[328] In 1866, Professor Frieze wrote to White requesting a $2,500 loan so that he could pay Professor Douglass when his mortgage came due. As collateral, Frieze offered the copyright on his *Virgil's Aeneid*.[329]

In a letter found in the Andrew D. White papers at the Bentley Historical Library, written in July 1861 from a Lizzie[330] in Ann Arbor to an Aimèe, Lizzie writes:

> Prof. Watson's affairs are much talked of just now. The impression is that he will be forced to resign his place, or else will be dismissed. He has been to Dr. Tappan in a most penitent mood, but I suppose his day of grace is past. He is dreadfully in debt and tradesmen of all kinds are crying out against him. Besides, his house is mortgaged for a large amount. It is a great pity for I suppose no young man ever began life with greater prospects than he did. Professors, Regents, and the Great Mogul[331] himself, all made a pet of him. His salary has always been good. I do not see how he can have run so deeply in debt.[332]

In September of 1862, most likely in an attempt to relieve himself of accumulated debts, Watson wrote to a gentleman in Saline, Michigan to offer his new home for sale for $5,000. The reasons Watson gave for his inducement to sell were explained as follows:

> My family consists simply of myself and my wife, and I do not wish to keep [sic] house. My duties at College re-commence October 1st and if I do not sell my place at once, I must furnish the house and live in it, as I prefer not to rent it. If therefore I can make an immediate sale I am willing to make such a sacrifice.[333]

Watson proceeded to convince the Regents to build a residence for him attached to the Observatory, thereby obviating the need for a residence provided at his own cost. (see Chapter 10)

Watson was distracted from his duties at the Observatory by his various business endeavors and his considerable worry over debts. During Brünnow's absence for a year in 1859-60 while serving as Associate Director of the Dudley Observatory in Albany, Watson had entire charge of the Detroit Observatory. Tappan's assessment of Watson's efforts during this time was that he contributed nothing:

> The year was spent by him in building operations, in life insurance agencies, in buying tax warrants, and in other speculations. I had a free and kind conversation with him on the subject, in which I urged him to desist, warned him that he would involve himself, and advised him to devote himself to his proper pursuits. He replied that he was bound to make money. I reminded him that all he had gained had been through science; and begged him to have confidence, that if he were faithful to science, science would take care of him. He replied that he could do both. As an instance of his successful speculations, he mentioned that he had bought a watch, and sold it again at a large advance. He soon became deeply involved, and lost his credit, as I had predicted.[334]

When Brünnow was called back to his post at Michigan by the Regents in 1860, at the urging of the Detroit donors, Watson was enraged and made his feelings known through an article published in the newspaper. Watson had wanted the post himself. His anger prompted him to fire off a string of letters to various life insurance agencies in New York and Connecticut asking to serve as an agent for the Ann Arbor vicinity.[335] This letter-writing campaign was successful, resulting in Watson's service as an actuary of the Mutual Life Insurance Company for several years.[336] Upon Brünnow's return, the Regents offered Watson the positions of Professor of Physics and Instructor in Mathematics to assist Dr. Devolson Wood, which Watson initially declined but eventually accepted. But, relations with Dr. Wood proved difficult. Watson made several attempts to locate another position, including volunteering to become a field officer in Lieutenant Gillis' regiment of Engineers and Mechanics in the U.S. Signal Service—with the stipulation that he be made a Major or Lieutenant Colonel.[337] Watson's credit problems worsened to the point that the Regents were on the verge of dismissing him. But, the action was delayed because of Watson's "earnest pleas, protestations and promises."[338]

Despite Watson's financial woes, he was generally well-liked, and respected as a scientist. In 1866, when Andrew D. White, as President of Cornell University, was contemplating the building of an observatory at Cornell, he wrote to Watson for advice. Watson recommended the telescopes of Alvan Clark and Sons, suggesting that he believed Clark was open to making a refractor of 24 inches clear aperture—twice the size of the Ann Arbor Fitz.[339] Of his vision of this Clark telescope Watson said "I would sit in sack cloth & ashes for a period ever so long for the privilege of using such an instrument."[340] After exchanging a few letters with White, Watson, always the opportunist, offered himself as the director. Coincidentally, Brünnow also wrote an unsolicited letter to White recommending C. H. F. Peters as the director of Cornell's planned observatory. As it happened, Cornell did not begin construction of its Fuertes Observatory until 1916, so neither Watson nor Peters got the job. Instead, Watson threw himself into yet another financial involvement, this time with Flowers' Gap Oil & Mining Co.

Watson was disgruntled with Michigan because he had only been able to raise about $6,000 of the $25,000 he wanted as an endowment to provide for the Observatory. Watson wrote to Andrew D. White "when I think of the embarassments [sic] by which I am surrounded, two grand instruments & no assistant, no fund for publication, I sometimes get almost disheartened. Still I hope on, and assure you I have only one aim in life & that is faithfully to do all that lies in my power to extend the science to which so much of my life has been devoted."[341] Seeing a potential opportunity at Cornell University, Watson later exclaimed to White "I have begged money until I am almost ashamed to make any more attacks . . . Sometimes, when ill success in my efforts here discouraged me, I have felt like applying to you to say a good word for me to the Trustees . . ."[342] Watson's lack of success with the Detroit donors may have resulted from the donors' disinclination to contribute to an endowment under Watson's management, because they were rankled over Tappan's dismissal, and perhaps over the disloyalty to Tappan that Watson had publicly expressed. They probably had little confidence in Watson's ability to manage money.

In June of 1864, Watson donated to the University Library a collection of his books on astronomy valued at over $900.[343] This action took place at the same time that Andrew D. White made a donation of books. Given Watson's financial troubles, one cannot help but speculate that Watson got the idea from White, with White paying Watson for the books to compensate him for the donation.

When Brünnow resigned in 1863 following Tappan's dismissal, Watson was able to realize his long-sought desire to be the director of the Detroit Observatory. Since 1860, when Brünnow returned to Michigan from Albany, Watson had been teaching Physics. But, seizing the opportunity created by Brünnow's departure, Watson nominated himself for the post and sought letters of recommendation from astronomers B. A. Gould, Elias Loomis,

William Chauvenet, Joseph Winlock, Benjamin Peirce, J. M. Gillis, and others. Not surprisingly, Watson had provided similar letters of support for positions friends were seeking. In particular, he provided a letter recommending the appointment of Gould at Harvard's observatory, when William Cranch Bond died. Efforts to get Gould appointed were widely considered to be bad form, however, because George Bond was the obvious successor to his father.[344]

Repeated criticism was expressed at Michigan about Watson because he did not admit visitors or students to the Observatory. The following note appeared in the bi-weekly student paper, *The Chronicle*, in May 1874:

> During the present week the juniors have been granted the privilege of making their long-wished for visit to the observatory. A passing glance at pale Luna and girdled Jupiter was allowed each man as his row slided along the seat, and then his only sight of the big telescope during his four year's course of study was over. Thinking we understand the connection of the Observatory with the University, a protest is just against such complete isolation of its advantages from those students who do not intend to spend life in formulating their way through space.[345]

While his teaching methods were both praised and criticized, he was popular with his students, though not necessarily for the right reasons. Watson's attention was paid only to those students who were eager to learn about astronomy, yet his charming discourse in lectures, his full voice, and his reputation for being an easy grader, attracted flocks of students. Watson was not attentive to the details of teaching: he is said to have given passing grades to an entire class, including one student who had died toward the beginning of the term.[346]

At least once, Watson attempted to take the short path in his research. After Brünnow left the University in 1863, Watson published in *Silliman's Journal* an article containing original material taken from Brünnow's "On the Determination of Orbits" without providing attribution. While Watson took the time to check with Brünnow to confirm that his interpolation of time distances of a comet could be performed in the way Watson proposed in the last six or seven lines of his article, Watson did not seek permission to publish formulae Brünnow had presented in his classroom lectures. Brünnow wrote to his former colleague at Michigan, Professor Edward P. Evans, about the incident, speculating that Watson plagiarized his work because he was desperate to show the University Regents some evidence of scientific productivity:

> I suppose Mr. Watson has thought it best to publish all the formulae, which I had given him in my lectures and which make up his entire paper excepting the last lines, because he knew, that they are not known much by American Astronomers, although they are in general use here in Europe. That he made it appear, as if they are his own inventions, was probably owing to his desire to have something to show to the Regents at the approaching crisis. If I had given him permission to publish the paper, I should have told him, that it would injure his reputation with European Astronomers at least, who would see, that he does not shrink from adorning himself with the merits of others.[347]

In striking contrast, Cleveland Abbe, who was educated at the Detroit Observatory by Brünnow and Watson, published a similar paper but called it "On the Improvement of the Elements of a Comet's Orbit: Brünnow's Method," and acknowledged as the source the lectures Brünnow had given in 1858.

To negative opinions expressed about him, Watson was indifferent, holding himself aloof from others and devoting his energies to his work. In a biographical memoir written after Watson's death, Professor George C. Comstock of the Washburn Observatory at the University of Wisconsin wrote of Watson:

> He had bitter enemies and they circulated reports, to the discredit of his personal character, which went uncontradicted and gained undeserved credence. It cannot be denied that a measure of truth attended many of these statements, but they were habitually distorted and magnified all out of proportion.[348]

That Watson was vain is indisputable. The notebooks he kept while a student at Michigan are repeatedly autographed by him every few pages. Sometimes the signature appears numerous times, as if he were practicing his autograph. In one case, he signs "James Craig Watson, Astronomer Royal," which is a designation given to only the most renowned astronomers of Europe. In a draft description of a telescope Watson proposed to make, he wrote:

> The Hon. James C. Watson, one of the greatest astronomers that this country has ever produced to whom immeasured devotion to science owes some of its greatest blessings. Astronomy under his patronage has reached a summit rarely attained.
>
> The Telescope which the Hon. James C. Watson, LL.D., F.R.S., F.A.S., &c &c &c proposes to make is of the Gregorian construction and will bear a magnifying power of 1200 Times! Great indeed!!! 1200! 1200![349]

Yet, Watson was well-liked for his pleasant personality and jovial nature. Throughout Watson's jotting books, and his letters, are scattered bits of whimsy, humor, and endearing use of words. His playful side comes through in a sketch he drew in his notebook to illustrate a mathematical problem he pondered. [Figure 11.6]

Figure 11.6. Illustration for a mathematical problem drawn in James C. Watson's 1855 jotting book shows his sense of humor. (Bentley Historical Library)

Titles and salary became measures of success and prestige in higher education in the late 1860s. Watson was elected to the National Academy of Sciences in 1867, and to the Royal Academy of Sciences in Catalina, Italy, in 1870. He received numerous honorary degrees, but he pursued them rather than receiving them as recognition bestowed by others. In 1869, Watson wrote to German astronomer, Dr. H. Bruhns, with a request to receive the title of Doctor. Bruhns responded that they would welcome his visit, but that he could receive the title Doctor from the University of Leipzig "in absentia" if he sent a copy of his *Theoretical Astronomy* and $50 (approximately $500 in 1998 dollars) "which one has to pay here for the promotion."[350] Watson received the Doctor of Philosophy degree from Leipzig in 1870; Yale University followed suit in 1871, again in response to Watson's letter of inquiry;[351] Columbia College conferred the Doctor of Law degree in 1877; and Watson was elected a member of the American Philosophical Society in the same year.

A disturbing comment was made about Watson by C. A. Young, a distinguished solar astronomer at Princeton University, in a letter written to George C. Comstock at the University of Wisconsin after Watson's death:

> ". . . his treatment of his wife was simply abominable. She was rather weak and querulous, . . . Of course I do not expect or desire this estimate of his character to go into the narrative of his life. I give it only to forestall any extravagant laudations which would vitiate the picture entirely for any one who knew the subject a little intimately. There is no need to expose his faults; but they should not be replaced by virtues he did not possess."

Young went further to say that Watson ". . . was one of the most energetic and able men I ever knew . . . extremely self-confident (but not perhaps more so than his abilities justified), selfish and unscrupulous in advancing his own interests."[352] Physically, he was described as being vigorous and healthy, but reaching 240 pounds toward the end of his life. He was, religiously, a fundamentalist, believing that it was impossible for a mathematician to be an atheist.[353]

An opportunity arose in 1878 for Watson to be director of the University of Wisconsin's new Washburn Observatory. Newspaper accounts suggest that Watson was wooed by both Wisconsin and Michigan "with an ardor nowadays reserved for football coaches."[354] Wisconsin succeeded in the recruitment, with Watson anxious for the opportunity to work with a much larger, new 15-inch Clark telescope. No doubt, Watson enjoyed all the attention. He moved into the residence that President Bascom had just vacated next to the Observatory, and energetically supervised the completion of the Observatory structure. [Figure 11.7] Characteristically, and probably due to his impatience, he undertook new initiatives using his own funds, such as the construction of a Students' Observatory and a Solar Observatory.[355] The Solar Observatory was of a highly unusual design, intended to be used to search for Watson's elusive intra-Mercurial planets. [Figure 11.8] It consisted of a twelve-inch underground tube

Figure 11.7. Residence in Madison (as it appeared in 1996) where James C. Watson and his wife lived next to Washburn Observatory at the University of Wisconsin. (Patricia S. Whitesell)

that extended up the hillside next to the main observatory, leading to a siderostat and long focus objective that could be used to examine the sky near the sun. Robert Bless notes in his history of the Washburn Observatory that "the notion that such a device can enhance visibility of objects near the sun dies hard and is still periodically revived."[356]

Watson died prematurely and unexpectedly at the age of forty-two on November 22, 1880 in Madison, Wisconsin. Professor Frieze wrote on December 9, 1880:

Watson had been at my house chatting in his usual delightful, entertaining manner only a few days before, full of hope, and in perfect health. No doubt, he sacrificed his life by imprudent exposure in his new house, occupied before it was sufficiently dry, and before the furnace work had been completed. Indeed, he was overtaken with the acute attack of cramps and chills which carried him away, while busy in the cellar about the furnace, at an early hour in the morning. Such mere carelessness is it that has taken away the most brilliant man we have ever raised up here.[357]

Another description of Watson's demise, written by his friend Professor Winchell of Michigan, appears to be more accurate because Watson wasn't building a residence at the time of his death. Winchell explains that Watson was working on the construction of his underground solar observatory "when a severe cold brought on peritonitis, which over-confidence in his physical powers permitted to reach a fatal stage before medical aid was summoned."[358] It may have been an ignored bout with appendicitis that finally overtook him.

Figure 11.8. Washburn Observatory circa 1891 at University of Wisconsin-Madison, with James C. Watson's small Solar Observatory in the foreground. (Mary Lea Shane Archives of the Lick Observatory, University of California-Santa Cruz)

Letters were received by the University of Michigan from around the country, mourning the tragic loss and speaking praises to the brilliant astronomer. Many gathered in Ann Arbor for the funeral when Watson was buried in Forest Hills Cemetery located near the Detroit Observatory. Andrew D. White wrote on Watson's death:

> I came to form a very high estimate of his genius and services. The breadth and depth of his knowledge and the hearty way in which he imparted it laid a great charm on me.[359]

In spite of Watson's predilection toward personal gain over scholarly pursuits, he expressed his devotion to science by attempting to perpetuate scientific excellence. He had amassed a considerable savings through his business activities (and very likely had established an insurance policy for himself because he was an insurance agent), leaving $15,000 (about $165,000 in 1998 dollars) through his will to the National Academy of Sciences, to which he had been elected a member in 1868. The funds were to be used for the calculation and publication of data on the asteroids he had discovered, and for the establishment of a gold medal recognizing a significant discovery or contribution to astronomical science, either in this country or abroad. The prize was awarded only once, to Watson's friend Benjamin Apthorp Gould.

Apparently, little was left for Watson's widow, Annette, and she returned to live with her parents in Dexter, Michigan, near Ann Arbor. She remained there for over twenty years, moving in 1904 to Ann Arbor where she lived out her life in a residence she had built at 212 Twelfth Street,[360] probably using funds from an inheritance left to her by her parents. At her death in 1916 at age 80, she was buried next to her husband in Forest Hills Cemetery.

Chapter 12

Brünnow Departs for the Dudley Observatory

During the year 1859-60, Franz Brünnow took his leave from Ann Arbor to work at the Dudley Observatory in Albany, New York [see Figure 8.5] as the Associate Director, having been called there by Ormsby M. Mitchel, the new director. The history of the Dudley Observatory is fascinating, yet tumultuous in its early years. It was established in 1852 at the initiation of the citizens of Albany, and with the financial backing of Mrs. Blandina Dudley in memory of her husband, former mayor of Albany and U.S. Senator, the Honorable Charles E. Dudley; and by Mr. Thomas W. Olcott, president of the Mechanics and Farmers' Bank of Albany. The objectives in establishing the Dudley Observatory were to promote science and national pride, to bring renown to the capital city, and as a first step in creating a proposed national university in Albany.

Ormsby M. Mitchel had been the invited speaker at the citizens meeting that was the inception of the plan, and he was asked to be the new observatory's director. But, Mitchel's time was occupied as director of the Cincinnati Observatory, and in spite of the difficulties in operating at long distance, he eventually did accept the appointment, though he filled it in absentia. The inauguration of the Dudley Observatory was timed to coincide with a meeting of the American Association for the Advancement of Science. Among the many distinguished guests at this event on August 14, 1856 were Tappan and Brünnow, accompanied by Charles W. Hackley and Edmund Andrews [see Chapter 5]. Professor Mitchel, overwhelmed with work at two observatories located so far apart, was forced to withdraw from his connection with the Dudley Observatory, and Professor Benjamin A. Gould of the U.S. Coast Survey, who was one of two men offered the director position at the Detroit Observatory before Brünnow accepted, stepped forward to be director of the Dudley Observatory. He had been one of the Dudley's principal donors, being a man of independent means.

The Dudley Observatory was equipped with instruments comparable to those at the Detroit Observatory, and by the same makers, among them a refracting telescope (1860) by Henry Fitz, Jr. and a meridian circle (1856) by Pistor & Martins. Clearly, there was competition between the Detroit and

Dudley observatories, and references in the literature to this relationship are numerous. Before long, the Scientific Council that governed the Dudley Observatory became impatient with the lack of progress Gould was making in setting up the Observatory. The Olcott meridian circle telescope had arrived in 1856, but it was still in its crate in 1859. Delays and other differences between Gould and the Scientific Council resulted in a lengthy "trial," complete with a written attack and an exhaustive, book-length defense. The final outcome was the removal of Dr. Gould.

The Scientific Council turned once again to Dr. Mitchel to assume the directorship. After discussion, he agreed, but deferred his arrival for one year due to his wife's ill health. Mitchel suggested to the Trustees that an assistant be appointed until he could relocate to Albany. Several individuals were considered, including C. H. F. Peters, a professor Sontag, and Franz Brünnow. It is not surprising that Mitchel thought of Brünnow as a candidate: Brünnow had already successfully set up nearly identical Pistor & Martins and Henry Fitz instruments at the Detroit Observatory.

All three potential candidates were Germans, which made the search process personally difficult for Mitchel. He confided in Thomas W. Olcott of the Dudley's Board of Trustees:

> I am afraid of foreigners in general & especially of those whom I know absolutely nothing by personal intercourse. This is of course strictly confidential, for I have no prejudices whatever against any nation, but simply feel that no foreigner can know as we do our country and our people & hence cannot so readily adapt themselves to our manners and customs.[361]

In spite of his difficulty accepting Germans, after meeting with Brünnow, Mitchel recommended to the Council that Franz Brünnow be invited to be the co-director, with full privileges including occupancy of the director's residence. Dudley Trustee Joel Rathbone, though he had full confidence in Brünnow, thought it unwise to have two directors.[362] Brünnow countered with a request to be appointed as Adjunct Director, but the title settled on was Associate Director, which Brünnow accepted.[363]

The appointment of Brünnow had been predicted by Gould several months in advance of the announcement,[364] suggesting that the indignities Brünnow was being made to endure at Michigan by certain Regents and the press were known beyond Ann Arbor. Yet, the reasons behind Brünnow's job change were not understood by all. G. P. Bond, director of the Cambridge Observatory at Harvard, wrote in response to Brünnow's announcement "I scarcely know whether to congratulate you or not, on your new appointment, but you would not have left your position at Ann Arbor without sufficient assurance of being supported in a better one at Albany."[365] In response, Brünnow explained that the offer from Mitchel had been unexpected.[366]

Beginning in 1858, President Tappan was experiencing difficulties with certain new members of the Board of Regents, particularly Regent Levi Bishop.

Under assumed names such as "Tresayle," Regent Bishop was writing acerbic letters to the *Detroit Free Press* attacking Tappan on numerous scores, representing Brünnow as incompetent and the Observatory a failure, and expressing regret that money had not been expended on a Law building instead. As a consequence of this hostile treatment, Brünnow accepted the appointment at the Dudley Observatory. In response, despite Regent Bishop's objection, the Regents passed resolutions which were complimentary of Brünnow, resulting in Brünnow's acceptance of a continuing, nominal directorship in Ann Arbor.[367] Bishop was furious, regarding the action as an effort to make the University "play second fiddle to a one-horse institution in Albany, not known in Europe, and to a Cincinnati institution which [had] always been inferior to Michigan."[368]

Brünnow was concerned that his motives in accepting the position at the Dudley Observatory might be misunderstood in Albany and in the scientific community. Gould had asked Brünnow to write a letter as a "witness" at Gould's controversial and complicated "trial," believing that the highly-regarded, fair-minded astronomer would aid his defense. As it turned out, Brünnow's letter provided the evidence that ultimately sealed Gould's demise. Brünnow wrote:

> . . . I feel sorry for you individually, but I must say *that you brought on all this difficulty by your own imprudence,* and by your treatment of Dr. Peters; and I must *entirely condemn* your course in this affair. I also regret that this has chilled the friendly feelings I had towards you since our first acquaintance.[369]

The Dudley Observatory was fortunate to have obtained the services of Franz Brünnow as Associate Director. His experience setting up the telescopes at Michigan made it possible to quickly set up the Olcott meridian circle and the other instruments, which had long sat in their crates. Soon, he had the Observatory in full operation. However, when Mitchel transferred his full attention to the Dudley Observatory in 1860, Brünnow apparently became dissatisfied with the arrangement: Gould reported that Brünnow left in a rage.[370] Perhaps Mitchel had been unable to contain his prejudice against Germans. In July 1860, after forty-two of the donors to the Detroit Observatory had prepared a petition to summon Brünnow back to his Director position at Michigan, the Board of Regents offered Brünnow his old job, and he readily accepted. He resigned his post in Albany on July 6, 1860.

Watson was livid about this turn of events, having had sole charge of the Detroit Observatory during Brünnow's absence. He expressed his disappointment to President Tappan, who urged him to set aside all such feelings and welcome the return of his faithful instructor. While it appeared to Tappan that Watson's unhappiness had waned, Watson responded by publishing an incendiary letter to the *Detroit Free Press* that attacked both Brünnow and Tappan's integrity. Although Watson subsequently apologized to the two, Tappan later said the letter "was probably the true exponent of [Watson's] feelings."[371]

Figure 13.1. Châlet Beauval at Vevey, Switzerland where the Tappans and Brünnows moved in 1880. Family members can be seen on the balconies. (Bentley Historical Library)

Chapter 13

Tappan and Brünnow Leave Michigan

Tappan began to experience opposition and ridicule at Michigan as early as 1853 when he returned from his trip to Europe to purchase astronomical instruments. Later, after a new Board of Regents was appointed in 1858, Regent Levi Bishop, and members of the press—in particular, W. F. Storey, editor and owner of the *Detroit Free Press*—took exception to Tappan's use of the title "Chancellor." They also criticized Tappan's eastern accent and pronunciations like "nevah" (never), which some misinterpreted to be pretension; the fact that Tappan sometimes drank wine with his meals; Tappan's attempts to "Prussianize" university education; the hiring of Franz Brünnow, a Prussian; the fact that the Tappans attended more than one church; Tappan's insistence in taking a non-sectarian approach to higher education; his support of the abolition of slavery; and his refusal to defer to the Regents' desire to micro-manage University business.

In Tappan's view, the role of the Board of Regents was "only to watch and defend, and not to interfere with the growth of what they have planted,"[372] a position he made clear in his inaugural address. Nonetheless, criticism of Tappan continued to escalate during his tenure until, through devious actions which took place in the final hours before the Regents' term of office ceased, they were successful in dismissing Tappan from the Presidency.[373]

On June 25, 1863, during the course of a routine Regents meeting, Regent Brown presented the following resolutions:

> WHEREAS, It is deemed expedient for the interests of the University that sundry changes be made in its officers and corps of Professors, therefore.
> *Resolved*, That Dr. Henry P. Tappan be and he is hereby removed from the offices of the University of Michigan and Professor of Philosophy therein.[374]

John Tappan [see Figure 1.6] was also dismissed from his position as University Librarian. And Brünnow, although not officially dismissed, was expected to resign given the circumstances—and he did.

Appeals and protestations were loud and widespread, coming from the many citizens of Ann Arbor, Detroit, and the State who admired and respected

Tappan. Tappan was stunned by the dismissal, but self-confident. He addressed the Board:

> This matter belongs to history: the pen of history is held by the hand of Almighty justice, and I fear not the record it will make of my conduct, whether private or public, in relation to the affairs of this University.[375]

The overthrow was planned in secrecy, as was later revealed in the reminiscence of Mr. George M. Lane, Class of 1853, at the dedication of the Tappan Memorial in 1914:

> Mrs. Lane and myself had come to Ann Arbor for the commencement exercises and that night, after we had retired to our room at the hotel, I was awakened by the hurried tramping of several people past our door. Apparently, they went into the room next to ours, and shortly afterwards, as the partition between the rooms was thin, I could not help overhearing much of what was said, and I very soon gathered that the people in the next room were Regents of the University and that they were determined to put Tappan out, at the meeting of the Board of Regents, the next day. I gathered, also, that the whole thing was to be kept secret, until that time. The next day I was invited to luncheon at the home of one of the Regents, and while at his table I ventured to ask him about the matter which I had overheard the night before. He looked at me in utter astonishment, and wanted to know what I knew about it. I told him I had certain very definite information. Upon which he begged me to keep silent, which I did, since I was not in a position to interfere. But that night's work has remained a blot upon the history of the University.[376]

As Colonel Isaac H. Elliott, Class of 1861, explained:

> [Tappan] was charged with being proud, and an aristocrat. The charge was true. He was proud, but not vain; pride is the rock that defies the storm; he was an aristocrat in mind and soul, he was kind and tender to the humble and did merciful things as stealthily as others committed crimes.[377]

Mr. Lane's final statement succinctly sums up the cause of the dismissal:

> Dr. Tappan was too big a man for the place, in that day.

After leaving Ann Arbor in 1863, the Tappans and Brünnows moved to Berlin, where they settled very comfortably in a pleasant part of the city. As was the custom in Germany, they occupied one floor of a house. Their third floor suite in a very large house included parlors, dining room, sleeping rooms, kitchen, etc.[378] The Tappans moved several times to various locations in Germany and Switzerland, but it was Switzerland that Tappan considered to be his spiritual home. He was captivated by Lausanne, which he thought made an ideal home where he could work in his study and look out the windows, walk in the garden, and climb the surrounding mountains for fantastic views. In 1880, an unexpected offer to purchase the Tappan's house in Basel was presented while they were away visiting in Thun. Within five weeks, they were settled into a new home in Vevey, Switzerland on the northeastern bank of

Lac Léman (Lake Geneva), not far from Lausanne. [Figures 13.1 and 13.2] Lac Léman became a refuge for "Romantics" after Jean Jacques Rousseau stayed there, and the droves of nature lovers who made pilgrimages to the area started a prosperous tourist industry which continues its popularity today.

Tappan described Mrs. Tappan's reaction to their new residence, called Châlet Beauval, as they approached by coach:

> In descending, the Coach passes in sight of Beau Val [sic], and handkerchiefs were waved from Coach & house windows. And what did Mrs. Tappan think of Beauval—what were the first impressions? She actually shed tears. She said Barbie in the letter she had written to her had exaggerated nothing, that all was far more beautiful and charming than she had imagined, although she anticipated a great deal.[379]

Of their new home, Tappan said:

> Vevey indeed is one of the most glorious and inviting spots on our globe . . . I have been struck during the past winter by the beauty and splendor of the mountain scenery of Switzerland in the season of frost & snow. . . . The upper vallies [sic] are filled with snow; the high peaks are capped with snow; and when the sunshine streams upon them there is a scene of beauty, grandeur, and dazzling splendour [sic] which no language can describe.[380]

M. NESTLER phot   VEVEY

Figure 13.2. "A walk on the Lakeshore at the foot of the lawn" is the inscription Julia Tappan wrote on the back of this photograph of the promenade along Lac Léman (Lake Geneva) at Châlet Beauval in Vevey, Switzerland where the Tappans and Brünnows enjoyed walking. (Bentley Historical Library)

Unfortunately, after Châlet Beauval was sold in 1905 by Rudolph Brünnow to the adjacent Grand Hôtel, it was demolished in 1928 and a restaurant was built on its foundations. During renovations in 1992, the old foundations and wine cellars were rediscovered.[381]

After relocating to Europe, Tappan never again took a formal position, but he continued to write, to speak, and to commune in learned circles. Tappan established a close friendship with William W. Murphy and his wife, which helped to alleviate some of the isolation the Tappans felt in Europe. Murphy was the American consul-general at Frankfort-on-Main from 1861-70, later serving in Germany as financial agent for various American railroad companies.

Tappan closely followed the events of the Civil War in European newspapers, writing often to friends about the things he read. In Berlin, Tappan presented, and later published, a memorial discourse on the death of Abraham Lincoln at the Dorotheen Church in Berlin on May 20, 1865, having been invited by N. B. Judd, the envoy and minister of the United States at the Court of Prussia, who was a personal friend of Lincoln.[382] Tappan was passionate about the United States. It was when Fort Sumter surrendered in 1861 at the outbreak of the Civil War that a patriotic fervor broke out in Ann Arbor. Tappan was called to address the public in the courthouse square. [Figure 13.3] His speech was described as powerful and inspiring, eloquent and earnest. He spoke with great emotion, and his speech was received with admiration. It was a speech long to be remembered: "In all Michigan's splendid history this was the great historic occasion. It was the same for the University. Both were from that hour for the Union and the war."[383]

The University of Minnesota offered Tappan the position of president in 1869, a position he considered but declined due to his advancing age and the fact that his family was happily settled in Europe. The University of Michigan extended an earnest invitation for him to attend the 1875 Commencement, which he declined with regret. It was at this time that the University resolved to erase from its history all criticisms of Tappan's leadership. Tappan could never fully accept this expression of apology, however. In his words, "a compliment does not repair a wrong" and "amends ought to be equal to the wrong."[384] But, Tappan looked back on his experience at Michigan with confidence that he could have been successful given different circumstances:

> My ambition was to make the University of Michigan equal to anything in the world: and it could have been done in time with proper men & measures.[385]

Tappan died on November 15, 1881 at his residence, Châlet Beauval, in Vevey, Switzerland at the age of seventy-six, and was buried there on the hillside overlooking Lake Geneva. Then-president of the University, Henry S. Frieze, in a Memorial Discourse he wrote in 1882 as a tribute to Tappan, described the fondness the University felt for its first president:

> Among his pupils he was quick to recognize ability and promise, and during his incumbency of the president's chair for eleven years he drew to himself the

Figure 13.3. Tappan's address in downtown Ann Arbor at the outbreak of the Civil War drew a crowd. (Bentley Historical Library)

esteem and affection of successive classes, and impressed every receptive and vigorous mind among them with something of his own strength and power. When he left he could justly take satisfaction in the knowledge that his pupils, while they respected him as a teacher, loved him also as a companion and friend, and bore him such reverence as children have for a father at once great in heart, broad in mind, and vigorous in intellect.[386]

Another passage, in a description of various university presidents, captures the essence of Tappan's personality and charm:

Tappan was of a warmer nature, judging men less severely, and seeing in life more of sunshine than of shadow.[387]

Frieze's final tribute to Tappan is a moving passage:

Thus lived and thus died Henry Philip Tappan, one of the most gifted men of our times; the Christian philosopher, the friend of Cousin, the lover of Plato; a cultivated scholar, a great educational leader; the first President of this University, and its true founder; whose work and memory are inseparable from its history; whose name shall live and be honored as long as the State and the University shall endure.[388]

Tappan's connection with animals and nature was a principal part of the essence of his character. While at Michigan, Tappan's German mastiffs, Leo and Buff—but particularly Leo—followed him everywhere, even to classes and church services. The dog pictured in the earliest known photograph of the Observatory may be Leo [see cover or Figure 14.1]. Of his connection with animals, Tappan said:

> My old dog Leo, who died the last summer I spent in Michigan, and whom I buried under a tree in my garden, often comes up before me when I sit alone and he seems to lay his head on my knee and to look up into my face with his gentle knowing eyes, and I feel how one feels when he recalls the memory of a departed friend. . . . To me, the relations between us and the domestic animals is a subject of deep interest and a home seems hardly complete without them.[389]

It is fitting that Leo is depicted at Tappan's side in the Tappan Memorial sculpture. Rendered by the artist Karl Bitter of New York, and installed in Alumni Memorial Hall in 1914, the sculpture depicts Tappan "linked with his dog Leo in perpetuity."[390] It was later placed in storage, but was retrieved in 1983 and placed in the lobby of Tappan Hall, where it remains today. [Figure 13.4]

Alumni reminiscences[391] of President Tappan, published after the Tappan Memorial sculpture was dedicated, are full of anecdotes involving Tappan's dogs and their interactions with people and with Professor Fasquelle's dog, Fido. But, perhaps the most touching evidence of Tappan's love of dogs appears in a letter written by Tappan to Professor Frieze from Vevey, Switzerland in September 1881, only two months before his death:

> I feel much touched by the tender care of Mrs. Frieze in planting foliage & flowers over my old dog's grave. And so do we all. Mrs. Brünnow sends her love to Mrs. Frieze & Miss Carrie, and desires me to thank her for the leaves from Leo's grave. I retain his photograph and I never look at it without experiencing in my heart a gush of tenderness that strengthens my faith that "love is indestructible" whether to man or beast.[392]

There was a dog at Dunsink Observatory, too, where Brünnow went after leaving Michigan to become Astronomer Royal of Ireland:

> Arrival: the *dog*, a huge specimen of black-faced pug, almost like a bulldog in power of fight, but very fat, and friendly with his master.[393]

Tappan remembers the dog affectionately:

> We had dogs at Dunsink and one at Willoughby Puy . . . He had most beautiful and expressive eyes & was very intelligent and had a great deal of human feeling. He was very fond of my daughter & I have known him to shed tears when she went away. He died and we buried him in our garden and planted a fir tree over his grave.

Figure 13.4. Tappan memorial bronze by Karl Bitter was installed in Alumni Memorial Hall in 1914 and later moved to Tappan Hall. Beside Tappan is his beloved dog, Leo, who was buried under a tree at the President's House. (Bentley Historical Library)

I now have a noble dog of the St. Bernard breed and born at the Hospice. He is about a year & a half old. He is very much attached to me & is my companion in all my walks, and lies under my table in my library & study.[394]

Henry P. Tappan was many things to many people. He was a philosopher, scholar, teacher, leader, and patriot, and he was a family man with a great love of animals, and mountains, and gardens. In their memorial tributes, his colleagues and acquaintances paid tribute to Tappan's commanding strength and achievements. Rebecca Lloyd Brünnow expressed her tribute by capturing in verse her father's softer side—the family man—in a touching poem she wrote following the death of her mother in 1884, in memory of her parents, Henry Philip and Julia Livingston Tappan. [Figure 13.5]

Figure 13.5. Henry Philip and Julia Livingston Tappan in old age, taken at Basel, Switzerland. (Bentley Historical Library)

In Memory of My Parents
by R. Ll. B.
November 15, 1881-August 21, 1884

In every nook and corner I am seeking
For the noble pair who once did grace this place.
My eyes are dim with constant bitter weeping;
I cannot find them, though I find their trace.

Their footprints on the sands of time implanted
We find at every turn, in every spot.
With their dear presence everything is haunted;
We feel them near, though they, alas, are not.

We hear their steps just coming round that turning
And think we hear *her* gentle, loving voice;
With joy we rise, *their* forms almost discerning,
Our grief is gone, our hearts again rejoice.

We stretch our hands in tender, eager greeting
To welcome them for whom our heartstrings yearn,
But never nearer comes that longed-for meeting—
To live without them we must sadly learn.

The library we enter softly, slowly;
Her seat still by *his* cozy study chair;
A something not of earth, benign and holy,
Here subtly seems to permeate the air.

It is that dim, mysterious, wondrous feeling,
Their presence from our midst forever gone,
Which through their old haunts cometh gently stealing
Like echoes of a well-beloved song.

O, n'er forgotten, though from hence departed,
Within our hearts we keep their honored place;
Though lonely, sad, we live on broken-hearted
Until we see them yonder, face to face:

Illuminated with light as opens wide the doors
Of our eternal heavenly home at last;
While streams of joy and peace divine immortal
O'erflow our souls, earth's woes forever past.

Châlet Beauval, Vevey, Switzerland, November 10, 1884.[395]

Figure 14.1. Earliest known photograph of the Detroit Observatory, circa 1858, taken by photographer T. D. Tooker. (Bentley Historical Library)

Figure 14.2. Detroit Observatory in August 1998 following its complete historic restoration. The sapling seen in Figure 14.1 above is now so large it obscures the dome. (Patricia S. Whitesell)

Epilogue

Perpetuation of a Legacy

Through the creation of memorials on campus and other efforts, attempts have been made to perpetuate the memory of the great contributions and achievements of Henry Philip Tappan at the University of Michigan. As Rudolph Brünnow so eloquently stated at the dedication of the Tappan Memorial in 1914, we must "forg[e] a living chain that will carry on the memory of Henry Philip Tappan to future generations."[396] It is intended that this book be a link in an ever-lengthening chain that documents and celebrates the life of Henry Philip Tappan and his many creations.

The Detroit Observatory is the University's closest link to its first President. The Detroit Observatory perpetuates Tappan's memory as a monument that was "a creation of [his] own"—a physical embodiment of his vision for higher education, and one of his first acts as President. The Observatory captures and provides a window in time back to the 1850s—to the University's first scientific laboratory, where the "Ann Arbor School of Astronomers" was born and significant scientific discoveries were made.

The 1997-98 historic restoration ensures that the Detroit Observatory will persist so that future generations can learn from the historic telescopes, gain an understanding of the context of the historical advancement of science in American higher education, and experience at close hand the path upon which the University of Michigan took its first steps toward becoming the research university it is today. [Figures 14.1 and 14.2]

Appendices

Appendix 1
Significance of the Detroit Observatory

Today, a step through the front door of the Detroit Observatory is the closest experience to travel back in time to the 1850s that is available on the University of Michigan campus, and perhaps the purest experience of 19th century astronomy available in this country. The Detroit Observatory is one of the University's most significant historical treasures. Constructed in 1854, it is the oldest basically unaltered building extant on the campus, second only in age to the President's House (1840), which has experienced numerous alterations and renovations.

The conception of the Detroit Observatory back in 1852 was the initial step taken by the University's first President, Henry Philip Tappan, toward creating a scientific course that would complement the traditional classical course of study. The Detroit Observatory, named to honor the city of its major bene-factors, thus became the University's first dedicated research laboratory. It was also the first observatory in Michigan (which was still a pioneer state in 1854), and second only to the Cincinnati Observatory (1843) in the Midwest.

Tappan recruited Franz Brünnow from Berlin as the first director of the Detroit Observatory. Brünnow, who was the first faculty member at Michigan to hold a Ph.D., introduced American students to rigorous European methods of astronomy, and he created the University's first scholarly journal, *Astronomical Notices*. His best student, James Craig Watson, discovered 22 asteroids (minor planets), and between them, they published dozens of important articles and ephemerides. The excellent reputation they earned for the Detroit Observatory prompted the term the "Ann Arbor School of Astronomers." It was said in those early days that Ann Arbor was *the* place to study astronomy.

In 1854, the small transitional Greek Revival building was isolated a half-mile from the campus, but today, its site has become the heart of the Medical Campus, placing it in the center of research and scholarly activities. Although the Observatory was virtually abandoned for several decades, it persisted largely due to the devotion of its many admirers. Today, the building speaks volumes about University and local history, and the early history of astronomy in this country.

In 1994, the Office of the Vice President for Research adopted the Observatory at the suggestion of the History & Traditions Committee. Private gifts made it possible to undertake a complete historic restoration that brought the building up to code and stabilized it to serve for another 150 years.

There are many compelling reasons to preserve the Detroit Observatory. Among the Observatory's credits, it is:

- The first observatory established in the State of Michigan, and second only to the Cincinnati Observatory (1843) in the Midwest;

- The oldest unaltered observatory in America that has its original instruments still intact, in their original mounts, and operational;

- The second oldest extant building at the University and the oldest campus building in its unaltered form;

- The University's first dedicated scientific research laboratory, led by Michigan's first faculty member to hold the Ph.D. degree, Franz Brünnow;

- The most important physical legacy of the University's early scientific preeminence;

- The site where numerous significant scientific discoveries were made, including 21 asteroids (minor planets) and two comets;

- The site where longitude was established in 1861;

- The training ground for many prominent astronomers of the 19th century;

- The home of the first scholarly journal published by the University: *Astronomical Notices,* created and edited by Franz Brünnow;

- Repository of what was the third largest refracting telescope in the world when it was made by Henry Fitz in 1857;

- Repository of the oldest, large objective telescope lens made by Henry Fitz that has not been refigured;

- Repository of the 1854 Pistor & Martins meridian circle telescope, the oldest intact instrument of its type in America, and the oldest Pistor & Martins meridian circle in its original mount in the entire world;

- The training ground for the first director of the U.S. Weather Bureau;

- One of the most perfectly preserved buildings of its era;

- Listed in the National Register of Historic Places.

To put into perspective one of the Detroit Observatory's special attributes—that it is the oldest extant observatory in this country that still retains its original telescopes in working condition in their original mounts—this is particularly noteworthy because it is the nature of astronomers to advance their science through the acquisition of bigger and more powerful telescopes. Accordingly, older telescopes are regularly replaced by new. Astronomers are constantly seeking new sites for observatories that are free of cloudy weather and interference from city lights and air pollution, and universities are loath to let buildings—or potential building sites—sit idle, preferring to renovate buildings for adapted uses or to raze buildings to make way for new structures.

Given these tendencies, the fact that the Detroit Observatory has persisted intact since 1854 verges on a miracle. The Observatory's solid construction, its small size and location on a hill that is undesirable as a building site for anything except a small building, efforts over the years of numerous preservationists and friends, and, ironically, just plain neglect over the years, have all contributed to the building's preservation.

Adding further significance to the Observatory is the refracting telescope located in the building's central dome, which retains its original objective lens, made in 1857 by Henry Fitz of New York. It appears that this lens is unique: it may be the only remaining large lens made by Fitz that has not been refigured by another telescope maker. Perhaps most significant of all, however, is the 1854 Pistor & Martins meridian circle telescope, still extant in the east wing of the Observatory. It is the oldest surviving Pistor & Martins meridian circle telescope in America, and one of only a few in the world. It is the oldest instrument of its vintage and type that is still in its original mount in the entire world.

With the completion of the Detroit Observatory's restoration, the building will again play a significant role in the academic life of the University. The Observatory will foster and promote academic, research, scholarly, and educational service functions related to the history of the University's early years. It will host symposia, a lecture series, conferences, service and training activities, and special events in a unique setting that will foster creative thinking through reflection on past challenges and achievements. The Observatory will once again serve as a laboratory for instruction in the use of astronomical instruments, and a museum of scientific equipment related to the Observatory's history will enrich instruction and interpret the building for visitors.

Many significant events are envisioned for the Observatory in the coming years. Perhaps most exciting will be the Observatory's 150th anniversary in the year 2004. Coincidentally, the year 2004 is the year in which the exceedingly rare transit of Venus (the passage of Venus across the face of the sun) will take place, an event which last occurred in 1882. There will be many things to celebrate, but perhaps the most important milestone to proclaim is the successful preservation of the building. May the Detroit Observatory stand forever so that its scientific, architectural, and scholarly history will be available to inspire and stimulate contemporary scholars, and continue to intrigue its many friends and admirers.

Appendix 2
Scientific Publications of Franz F. E. Brünnow
and James C. Watson while at the Detroit Observatory[397]

Franz Brünnow

Tables of Flora, with reference to the perturbations by Jupiter and Saturn, published by the Royal Academy of Berlin, 1855

Tables of Victoria, with the perturbations by Jupiter and Saturn, published by authority of the Board of Regents of the University of Michigan, 1858

In the *Astronomische Nachrichten* published at Altona, in Denmark
General perturbations of Victoria by Jupiter and Saturn, and ephemeris for the opposition, 1857
Ephemeris of Victoria for the opposition, 1858-59

In the *Astronomical Journal* published at Albany, New York
Observations of Flora
Elements of Clio
Observations of Metis and Flora
Observations of Calliope, Thalia, Masea, Iia, Hebe and the Comets of 1858 I, 1858 II, and 1858 III

In *Astronomical Notices,* published in Ann Arbor and Albany, New York
The General Perturbations and Elliptical Elements of Vestia
Observations and Elements of Donati's Comet
On the Problem of the Shortest Twilight
On the Calculation of the True Anomaly in Ellipses and Hyperbolas of Great Eccentricity
Stars on the Parallels of Alexandria and the 55th Asteroid
On the Periodical Comet of DeVico
Elements and Ephemeris of Pandora
Stars on the Parallel of Pandora
Sweeping Ephemeredes for DeVico's Comet in 1860
Remarks on Donati's Comet
Observations of Asteroids
Observations of Asteroids, made at the Dudley Observatory, with the Olcott meridian circle
Examination of the Division of the Ann Arbor meridian circle
Observations of Proserpine, made with the Olcott meridian circle, at the Dudley Observatory
Observations of Asteroids, made at the Dudley Observatory

On the Latitude of the Dudley Observatory
On a Magnetic Break Circuit
On the Longitude of the Dudley Observatory
Observations of Asteroids made with the Walker meridian circle
    at Ann Arbor
On the Great Comet
On the Determination of the Flexure of Astronomical Instruments
Determination of the Longitude of Ann Arbor by the Telegraphic Method

James C. Watson

A Popular Treatise on Comets, Philadelphia, 1860

*Theoretical Astronomy relating to the Motions of the Heavenly Bodies around the
    Sun in accordance with the Law of Universal Gravitation,* Philadelphia, 1868

In the *Astronomische Nachrichten* published at Altona, in Denmark
    Observations of the Fourth Comet of 1857
    Elements and Ephemeris of the Fifth Comet of 1858
    Elements of the Third Comet of 1858
    Observations, Elements and Ephemeris of the Fifth Asteroid
    Investigation of the Orbit of Eurynome
    Discovery of an Intra-Mercurial Planet

In the *Astronomical Journal* published at Albany, New York
    Elements of the First Comet of 1857
    Elements of Ariadne
    Elements and Ephemeris of the Fourth Comet of 1857
    Observations of Clio
    Elements and Ephemeris of the Sixth Comet of 1857
    Elements of the First Comet of 1858
    Elliptic Elements of the First Comet of 1858
    Observations of the Comets of 1857 IV, and 1857 V, and the Asteroids
        Letitia, Virginia, Hestia, Aglaia and Calliope
    Elements and Ephemera of Calypso
    Elements and Ephemera of the Third Comet of 1858
    Observations of the Comets of 1858 I, 1858 II, and 1858 III, and the
        Asteroids Europa, Nemausa and Atlanta
    On the Orbit of Hestia
    Elements and Ephemeris of the Fifth Comet of 1858
    Observations, Elements and Ephemeris of the Fifty-fifth Asteroid
    Elliptic Elements of the Fifth Comet of 1858

In *Astronomical Notices,* published at Ann Arbor and Albany, New York
  Corrected Observations of Comets and Asteroids, 1858
  Observations of Encke's Comet, 1858
  Ephemeris of Hestia for the Opposition of 1859
  Observations of Pandora and Hestia
  Observations of Virginia
  Discovery of a Comet
  Elements and Ephemeris of the First Comet of 1859
  Observations of the First Comet of 1859
  On the Orbit of Pandora
  On the Orbit of Hestia

Other publications:
  On the Extraction of Roots, *Michigan School Journal,* 1859
  Tables for the calculation of simple or compound interest and discount and averaging of accounts, Ann Arbor, 1878
  American watches: an extract from the Report on Horology at the International Exhibition at Philadelphia, 1876, 1877

Appendix 3
List of the Asteroids
Discovered by James C. Watson[398]

No.	Name	Date of Discovery
79	Eurynome	1863, Sept. 14
93	Minerva	1867, Aug. 24
94	Aurora	1867, Sept. 6
100	Hekate	1868, July 11
101	Helena	1868, Aug. 15
103	Hera	1868, Sept. 7
104	Klymene	1868, Sept. 13
105	Artemis	1868, Sept. 16
106	Dione	1868, Oct. 10
115	Thyra	1871, Aug. 6
119	Althaea	1872, April 3
121	Hermione	1872, May 12
128	Nemesis	1872, Nov. 25
132	Æthra	1873, June 13
133	Cyrene	1873, Aug. 16
139	Juewa*	1874, Oct. 10
150	Nuwa	1875, Oct. 18
161	Athor	1876, April 16
168	Sibylla	1876, Sept. 28
174	Phaedra	1877, Sept. 2
175	Andromache	1877, Oct. 1
179	Klytaemnestra	1877, Nov. 11

*All asteroids were discovered at the Detroit Observatory except Juewa, which was discovered while Watson was in Pekin, China. To honor his hosts, Watson asked them to name the new planet.

Comets Discovered by John M. Schaeberle
at the Detroit Observatory[399]

1880 II	April 7, 1880
1881 IV	July 14, 1881

Appendix 4
Chronological List of Directors
of the Detroit Observatory

Appointed

1854	Franz Friedrich Ernst Brünnow
1864	James Craig Watson
1879	Mark Walrod Harrington
1892	Asaph Hall, Jr.
1905	William Joseph Hussey
1926	Ralph Hamilton Curtiss (acting)
1927	Ralph Hamilton Curtiss
1929	W. Carl Rufus (acting)
1930	Heber Doust Curtis
1941	W. Carl Rufus
1945	Alan D. Maxwell (acting)
1946	Leo Goldberg
1960	Freeman D. Miller (acting)
1961	Orren C. Mohler
1970	W. Albert Hiltner

Appendix 5
Some Notable Astronomers Educated
at the Detroit Observatory, 1854-1910[400]

Asaph Hall, Sr., Postgraduate, 1855; LL.D. Yale, 1878. Assistant Observer, Harvard Observatory, 1857-62; Professor of Mathematics, U.S. Naval Observatory, 1863- . Discoverer of Deimos and Phobos, the two satellites of Mars; awarded gold medal of the British Astronomical Society in 1878. His discussion of the observations made is considered by astronomers a masterpiece.

James Craig Watson, A.B. 1857, M.A. 1860; Ph.D. Leipzig, 1870 and Yale, 1871; LL.D. Columbia, 1877. Assistant Observer, University of Michigan, 1858-59; Professor of Astronomy and Instructor in Mathematics, 1860-63; Professor of Physics and Instructor of Mathematics, 1860-63; Director, Detroit Observatory, 1863-79; Professor of Astronomy and Director, Washburn Observatory, University of Wisconsin, 1879-80. Gold medal awarded by the French Academy of Sciences for his discovery of 22 minor planets (see list in Appendix 3).

Cleveland Abbe, B.A. New York Free Academy, 1857; M.A. University of Michigan, 1860. Instructor in Physics and Civil Engineering, 1859-60; Aide to Dr. Gould in the U.S. Coast Survey, 1860-64; Aussur Elet Massigre of Imperial Observatory at Pulkova, Russia, 1864-66; Aide in U.S. Naval Observatory, 1861-68; Director, Cincinnati Observatory, 1868-71; Meteorologist, Weather Service, U.S. Army Signal Corps, 1871-1916. Trained in astronomy by Brünnow and in engineering by DeVolson Wood. Inaugurated a system of Telegraphic Weather Reports (1869) for the use of the Cincinnati Chamber of Commerce; plan adopted by the Government in 1871 and used as a basis for the War Department's Weather Prognostications. Published 73 papers on mathematical and meteorological topics in American and foreign journals.

William W. Payne, 1863-64. Director, Goodsell Observatory, Carleton College, 1871-1908; Director, Elgin Observatory, 1908- ; founder, publisher and editor, *Sidereal Messenger* and *Popular Astronomy*.

George Benjamin Merriman, B.A. Ohio Wesleyan, 1863; M.A. on examination, 1864; M.A. Flint, 1863. Assistant in U.S. Naval Astronomical Expeditions in Chile, 1864-66; Assistant in Detroit Observatory and Assistant Prof. Mathematics, 1866-71; Adjunct Prof. of Physics. Engaged in preparing Star Tables for use in preparing the Nautical Almanac at the U.S. Naval Observatory, 1871-74; Prof. Math., Albion College, 1875-77; Prof. of Mathematics and Astronomy, Rutgers University, 1877-91; Prof. of Mathematics and Astronomy, Middlebury College, Vermont, 1891-94; Prof. of Mathematics and

Astronomy, Lawrence University, Wisconsin, 1894-99; Assistant on the Nautical Almanac, U.S. Naval Observatory, 1899. Participated in the Eclipse Party to Burlington, Iowa in August 1869 (see photograph Chapter 11, Figure 11.3).

Mark Walrod Harrington, A.B. Northwestern University, 1865; M.A. University of Michigan, 1868; Ph.D. Leipzig, 1877; LL.D. (hon.) U. Michigan, 1894. Assistant Curator of Museum of Geology, Zoology and Botany, U. Michigan, 1868-69; Instructor in Math. and Asst. Curator, 1870-72; Acting Astronomical Aide of U.S. Survey of Coast of Alaska, 1870-71; Instructor in Geology, Zoology, and Botany, U. Michigan, 1872-73; Student in Leipzig, 1876-77; Prof. of Astronomy and Mathematics in the Cadet School of the Chinese Foreign Office, Pekin, China, 1877-78; Prof. of Natural Science, Louisiana State University, 1878; Prof. of Astronomy and Director, Detroit Observatory, 1879-92; Chief, U.S. Weather Bureau, 1891-95; President, University of Washington, 1896-98.

William F. McK. Ritter, A.B. 1871, A.M. 1874. Assistant, U.S. Naval Observatory and Nautical Almanac Office.

Robert Simpson Woodward, C.E. 1872, LL.D. (hon.) 1912. Astronomer, U.S. Transit of Venus Commission, 1882-84, U.S. Geological Survey, 1884-90, and U.S. Coast and Geodetic Survey 1890-93; President, Carnegie Commission of Washington, 1904-20.

Monroe B. Snyder, A.B. 1872, A.M. 1875. Director, Philadelphia High School Observatory.

Otto Julius Klotz, C.E. 1872, Sc.D. (hon.) 1913. Astronomer, Alaska Boundary Survey, 1893-94; Director, Dominion Observatory, Ottawa, Canada, 1917- .

Charles Leander Doolittle, C.E. 1874, Sc.D. (hon.) 1897. Astronomer, U.S. Boundary Survey, 1873-75; Director, Sayre Observatory, Lehigh University, 1875-95; Director, Flower Observatory, University of Pennsylvania, 1895-1912.

John Martin Schaeberle, C.E. 1876, M.S. (hon.) 1893. Assisted Watson at Detroit Observatory in preparation of "Lunar Tables," 1877-78; Assistant in the Detroit Observatory, 1877-78; Acting Professor of Astronomy, University of Michigan, 1878-88; Astronomer, Lick Observatory, 1889-98; Acting Director, Lick Observatory, 1897-98; discovered two comets, 1880 and 1881.

George Cary Comstock, Ph.B. 1877, Sc.D. (hon.) 1907. Professor of Astronomy, Ohio State University, 1885-87; Director, Washburn Observatory, University of Wisconsin, 1889-1922; Dean, Graduate School, University of Wisconsin, 1906-20.

Mary E. Byrd, A.B. 1878. Director, Smith College Observatory, 1887-1906.

Edward Israel, A.B. 1881. Astronomer, Greeley Polar Expedition.

William Wallace Campbell, B.S. (C.E.) 1886, M.S. (hon.) 1899, Sc.D. 1905. Instructor of Astronomy, University of Michigan, 1888-91; Astronomer, Lick

Observatory, 1891-1901; Director of Lick Observatory, 1901-30; President, University of California, 1923-30; President, National Academy of Sciences, 1931-35.

Arman Otto Leuschner, A.B. 1888, Sc.D. (hon.) 1913. Director, Students' Observatory, University of California, 1898-1938; Dean, Graduate School, 1913.

William Joseph Hussey, B.S. (C.E.) 1889. Assistant, Nautical Almanac, Washington, 1889; Instructor in Mathematics, U. Michigan, 1889-91; Instructor in Astronomy, 1891; Assistant Prof. of Astronomy, Leland Stanford Junior College, 1892-94; Professor, 1894-96; Astronomer, Lick Observatory, 1896-1905; Director, Detroit Observatory, 1905-26; Director, LaPlata Observatory, Argentina, 1911-17.

A. L. Colton, Ph.B. 1889, A.B. 1890, A.M. 1898. Assistant Astronomer, Lick Observatory, 1892-97.

Herbert L. Rice, 1889-91. Assistant, Nautical Almanac Office, 1892-1902; Astronomer, Naval Observatory, 1902-07; Professor of Mathematics, U.S. Naval Observatory, 1907- .

James Robertson, B.S. 1891. Assistant, Nautical Almanac Office, 1892-1929; Director, Nautical Almanac Office, 1929-36.

Heber D. Curtis, A.B. 1892. Astronomer, Lick Observatory, 1892-1929; Director, Detroit Observatory, 1930-41.

J. C. Hammond, B.S. (M.E.) 1894. Astronomer, U.S. Naval Observatory.

Walter M. Hamilton, A.B. 1894, A.M. 1896. Assistant, Nautical Almanac Office.

Sidney Dean Townley, Ph.D. 1897. International Latitude Observer at Ukiah; Professor of Astronomy, Stanford University.

O. M. LeLand, B.S. (C.E.) 1900, C.E. 1920. Professor of Astronomy, Cornell University; Astronomer, Alaska Boundary Survey; Dean, Department of Engineering, University of Minnesota.

Harriet Bigelow, Ph.D. 1904. Director, Smith College Observatory.

Frank D. Urie, A.B. 1910. Astronomer in Elgin Observatory.

Appendix 6
Detroit Observatory Chronology

1852

October	Henry P. Tappan visits Ann Arbor
December	Tappan inaugurated as first President of the University
	Henry Walker organizes a fundraising meeting at the Michigan Exchange In Detroit

1853

February	Tappan and Walker order the Fitz refracting telescope in NYC
	Tappan travels to Europe by way of New York
	Tappan tours the observatories of Europe
July	Tappan meets Professor Encke and Dr. Franz Brünnow
July 15th	Tappan orders a meridian circle telescope from Pistor & Martins and a sidereal clock from M. Tiede
	Land is obtained for a site to build the Observatory, and for a road to the building; Samuel Pettibone surveys the site
August 5th	Ground is broken and construction begins
November	Tappan returns to Ann Arbor and orders two collimating telescopes of Pistor & Martins for $375
	Tappan offers director position to Prof. W. A. Norton of Yale College, and then to Dr. B. A. Gould of Boston, but they both decline. Position is then offered to Dr. Franz Brünnow, of Berlin, who accepts

1854

June	Brünnow's appointment as Director is approved by the Regents
	Delivery of the refracting telescope is expected
July	Brünnow arrives in Ann Arbor
	Fitz refractor arrives but must be dismounted to make alterations
	Fitz 4-inch comet seeker obtained
September	Meridian circle telescope arrives in Ann Arbor
	Sidereal clock (No. 125) by M. Tiede arrives
	Observatory completed

1855

	Observatory is enclosed by a new fence
	New road is constructed to the Observatory
	Well and drain are dug
	Brünnow publishes his *Tables of Flora*
April	Temporary refracting telescope mounted

June	Board of Regents tours the completed Observatory
August	Samuel P. Avery introduces Tappan to Cropsey
September	Jasper F. Cropsey visits Ann Arbor; paints the Detroit Observatory and a campus landscape
	R. Leggett prepares engravings of the Cropsey paintings for use in 1855-56 *Catalogue* and on University letterhead
November	Professor Frieze visits Brünnow's parents in Berlin
December	Replacement refractor arrives, but it is found to be defective and is returned within a few months; new contract is negotiated with Fitz for a completely new instrument

1856	Observatory building is stuccoed
March	Walker organizes a second fundraising meeting in Detroit, at the National Hotel
August	Tappan travels to Detroit with Hackley, Andrews and Brünnow for the American Association for the Advancement of Education meeting where Tappan delivers an address: "John Milton on Education"; they travel on to Albany for the dedication of the Dudley Observatory, and then meet with Henry Fitz in New York
	George Saunders is paid $115 to furnish a stone pier for the Fitz refractor
	A stove is added to the west wing of the Observatory
	Tappan commissions Cropsey to do a painting of Genevieve
	Cropsey moves to England

1857	
March	Franz Brünnow marries Rebecca Lloyd Tappan; they take an 8-month leave to visit Brünnow's family in Berlin
	James Craig Watson, Brünnow's student, graduates
	Cleveland Abbe enrolls at the University
	Embankment surrounding Observatory is sodded
	Wood house is constructed
July	First telegraphic cable is laid across the Detroit River
October	Brünnow returns from leave in Europe
November	New Fitz refractor arrives and is mounted in the dome

1858	Rudolph Brünnow is born
	Brünnow publishes his *Tables of Victoria*
	Brünnow begins publishing the University's first scholarly journal, *Astronomical Notices*
	Detroit donors ask Brünnow to investigate how to telegraph the time from the Observatory to Detroit

1859   Brünnow accepts position as Associate Director of the Dudley
            Observatory in Albany, NY

1860   Regents call Brünnow back to Ann Arbor in response to an
            appeal by the Detroit donors
            Brünnow gets approval to purchase a chronograph from the
            Bonds for $360
            Brünnow seeks Regental approval to purchase 2½ acres of land
            to the west of the Observatory to build a residence for
            himself, although granted, a residence is never built
            Part I of Brünnow's *Lehrbuch der spärischen Astronomie* is
            translated into English
            Watson marries Annette Helena Waite of Dexter, MI and builds
            a house on South University Avenue
October   Brünnow reports that a telegraphic connection was established
            with the Lake Survey in Detroit

1861   Telegraphic connection is made with the Litchfield
            Observatory at Hamilton College, Clinton, NY; longitude
            of the Detroit Observatory is established
            Time service was probably established in this year

1862   Watson offers his house for sale
March   At Brünnow's request, the Regents form a committee to
            consider the expediency of relocating the Observatory
            Statue of Nydia is purchased
            Cropsey is honored by Queen Victoria

1863   Henry Fitz, Jr. dies
June 13th   Tappan writes away to get Brünnow's passport renewed
            because Brünnow intends to travel alone to Germany on July 4th
June 25th   Tappan and his son, John, are dismissed by the Regents;
            Brünnow resigns shortly thereafter
            James Craig Watson, Brünnow's student, becomes Director
            Andrew D. White acquires Cropsey paintings and later returns
            them to the University in 1890
September   Watson discovers his first asteroid, named Eurynome

1864
June   Repairs made and new poles obtained for telegraph connection
            to the Observatory
            Telegraphic connection established with the United States
            Lake Survey to determine the longitude of Green Bay, WI

1865   Brünnow appointed Astronomer Royal of the Dunsink
                    Observatory, Dublin, Ireland
                Brünnow completes the translation of his *Lehrbuch der
                    spärischen Astronomie* into English
                Wooden columns are removed from the dome room, and
                    another course of bricks is added to strengthen the exterior
                    wall of the dome
August   Watson proposes that the Observatory be rebuilt on the
                    southeast corner of the University Grounds
                Jonas Phelps, who made the mount for the Fitz refractor, dies

1866   East wing over Meridian Room is re-roofed and repairs are
                    made to the meridian circle telescope

1867   Watson discovers asteroids Minerva, Aurora
                Detroit honors Tappan by naming a new school The Tappan
                    School

1868   A director's residence is added to the west side of the
                    Observatory
                City of Ann Arbor improves the streets around the Observatory
                Watson publishes his *Theoretical Astronomy*
February   Tappan writes a letter in defense of Samuel F. B. Morse's claim
                    to have invented the electro-magnetic telegraph
                Watson discovers asteroids Hekate, Helena (named after his
                    wife), Hera, Klymene, Artemis, Dione

1869   Brünnow is elected a Fellow of the Royal Astronomical Society
                Tappan declines the presidency of the University of Minnesota
                Watson observes a total solar eclipse at Mt. Pleasant, Iowa
                Longitude established directly with the Harvard Observatory
August   Bookcases are added at the Observatory, and a case for the
                    chronograph

1870   Watson works on zone observations for the U.S. Coast Survey
                Watson observes an eclipse in Sicily

1871   Watson discovers asteroid Thyra
                Brünnow expands and updates *Brinkley's Astronomy*

1872   Watson discovers asteroids Althaea, Hermione, Nemesis

1873   Watson discovers asteroids Æthra, Cyrene

1874	Brünnow retires from Dunsink Observatory due to failing eyesight
	Watson takes a one year leave for the transit of Venus in Pekin, China, and discovers an asteroid the Chinese named Juewa
	Watson travels from China to India, Egypt, and Europe
	Watson works on geodetic survey of Egypt
	Watson decorated as a Knight Commander of the Imperial Order of the Medjudieh of Turkey and Egypt, and takes a houseboat trip up the Nile
October	Tappan is invited to attend commencement, but declines
1875	Watson discovers asteroid Nuwa
1876	Watson discovers asteroids Athor, Sibylla
1877	Watson discovers asteroids Phaedra, Andromache, Klytaemnestra
1878	U.S. Government station is constructed by the U.S. Government on the grounds behind the Detroit Observatory to observe the transit of Mercury in May 1879; it becomes the Students' Observatory
	Watson claims to have discovered planet Vulcan while observing a total solar eclipse at Rawlins, Wyoming
1879	Watson appointed first Director of the Washburn Observatory at Madison, Wisconsin
	Mark W. Harrington, Class of 1868, becomes Director of the Observatory
	Meteorological instruments are purchased for the Observatory, including a Hough's barograph, Hough's thermometer, and an anemograph of the Lte. Gibbon's pattern
	U.S. Signal Service loans weather instruments: standard thermometer, psychrometer, terrestrial-radiation thermometer, solar-radiation thermometer
	Law professorship created in Tappan's name
1880	Watson dies suddenly in Madison, Wisconsin; buried in Ann Arbor at Forest Hills Cemetery near the Observatory
	Telescopes are purchased for the Students' Observatory to replace instruments recalled by the Government: a 6-inch equatorial refractor, a 3-inch transit instrument with zenith telescope attachment by Fauth & Company but with Clark optics, and a chronometer by Elgin

J. M. Schaeberle, Assistant in the Observatory, discovers a comet
The Brünnows move from Basel to Vevey with the Tappans to
    their new home, Châlet Beauval

1881   Tappan dies in Vevey, Switzerland
    Schaeberle discovers another comet
    Meteorologic observations commence; furnished to the
        State Board of Health until 1905
    Harrington takes a leave of absence to observe on the
        Pacific coast

1882   Transit of Venus observations take place, using the Students'
        Observatory
    Observations of the great comet of 1882
    Rudolph Brünnow receives his doctorate from Strassburg

1883   Furnace added to the Director's residence
    Coal house on the Observatory grounds is converted to a barn

1884   Harrington is founder of the *American Meteorological Journal*
        (the first and only such journal in the U.S.)
    Julia Tappan dies at Vevey, Switzerland
    Rebecca Tappan Brünnow publishes a poem about her parents
    Rudolph Brünnow appointed Professor of Semitic Philology at
        University of Heidelberg

1887   Dr. Ludovic Estes determines the latitude of the Observatory

1888   Schaeberle departs for Lick Observatory
    William W. Campbell, trained by Schaeberle, publishes his
        Elements of Practical Astronomy

1889   The Brünnows move to Heidelberg to be with Rudolph

1890   Dome converted: new shutter by Warner & Swasey; dome
        cannon balls replaced with railroad-style wheels; new
        observer's chair of the Burnam-Hough pattern
    Lighting updated on the meridian circle telescope
    Andrew D. White returns the Cropsey paintings to the University

1891
July   Harrington becomes first Chief of the U.S. Weather Bureau
August   Franz Brünnow dies

October   Professor Harrington donates a Zöllner Photometer to the
                 Observatory

1892   Hussey departs for Leland Stanford Junior University
                 Asaph Hall, Jr. appointed as Director
                 Dr. Richard H. Bull, the Observatory's architect, dies

1893   E. Howard clock No. 413 obtained to keep sidereal time
                 Repsold micrometer obtained for the meridian circle
                 Chronograph by Saegmüller obtained
                 Asaph Hall receives Regental approval to purchase a sextant
                 Hall dismounts the Fitz refractor for cleaning
                 Hall makes repairs to the meridian circle
July   Rebecca Tappan Brünnow dies

1894   Rudolph Brünnow marries Marguerite Beckwith
                 Tappan Hall, the new recitation building on campus, is named
                 in Tappan's honor

1896   Harrington serves as President of the University of Washington
                 Harrington gets approval to add bathroom to Observatory

1898   Harrington returns to the U.S. Weather Bureau as Director at
                 San Juan, Puerto Rico, returning to New York after 6 months
                 Harrington retires due to failing physical and mental health

1900   Jasper F. Cropsey dies

1902   Alvan Clark & Sons add a spring to the meridian circle lens cell
                 Sextant purchased for $150

1903   Astronomer Cleveland Abbe, Class of 1860, writes in support
                 of moving the Observatory to a better location
                 Surveyor's transit purchased for $375

1904   Hall publishes *Determination of the Aberration Constant*

1905   Hall resigns to return to the U.S. Naval Observatory
                 William J. Hussey appointed as Director
                 Director's residence expanded and improved
                 Observatory Shop established
                 Weather observations sent to the U.S. Weather Bureau
                 Hussey travels to Egypt to observe solar eclipse

1906	Observatory Shop established; staff hired
January	Bathroom added to original observatory
August	37½-inch mirror ordered from Brashear for a new reflecting telescope; delivered in December

1907
Fitz refractor converted: steel tube replaces wooden tube, driving clock added, 3½-inch finder, etc.
New Warner & Swasey micrometer obtained for the Fitz refractor
New observer's chair obtained—lighter and more convenient
Drainage improvements made to the Observatory
Observatory becomes a station of the U.S. Weather Bureau
Courses added in modern astronomy and astrophysics, history of astronomy, and variable stars
Rudolph Brünnow's wife dies; he moves to Princeton, NJ

1908
Students' Observatory moved 300 feet west
West wing of basement in original building is expanded
1908 addition is constructed, attached to the original building at the east
New comet seeker telescope constructed by Henry Colliau in the Observatory Shop, with Brashear 4½-inch lens; mounted on top of Director's residence
Catwalk added from original building to roof of residence; another to the 1908 addition
Original comet seeker used as a finder on the 37½-inch reflecting telescope
Students' refractor receives a new driving clock, worm and worm wheel, and electric slow motion
Camera provided for the students' refractor
Encroachment on the Observatory site was stopped after bulldozers began grading the land to the west as a women's athletic field
Robert Lamont donates funds for refracting telescope to be used at the Lamont-Hussey Observatory in Bloemfontain, South Africa
Harrington, missing for years, discovered homeless in Newark and is placed in a sanitarium

1909
Seismographs added to the clock room in the 1908 addition
Single-prism spectrograph by Brashear obtained for the 37½-inch reflecting telescope

1910	40-foot dome for the 1908 addition is constructed and installed by the Russell Wheel and Foundry Co., Detroit

1910   40-foot dome for the 1908 addition is constructed and installed
         by the Russell Wheel and Foundry Co., Detroit

1911   37½-inch reflecting telescope for spectrographic work is
         completed
         Hussey appointed Director at LaPlata Observatory in
         Argentina, in addition to directorship at Michigan
         Ralph H. Curtiss appointed Assistant Director
         Observatory grounds expanded to the east by gift of 26 acres
         from R. P. Lamont, Civil Engineering Class of 1891
         Howard clock No. 413 moved to the 1908 addition
         37½-inch reflecting telescope completed

1912   *Publications of the Observatory of the University of Michigan* series
         is launched; Vol. I, Part I, is published
         Plan developed to drain the Cat-Hole and build a University
         Power Plant; site called Huddy Hill considered for the
         Observatory

1914   Dedication of the Tappan Memorial; Rudolph and Eric
         Brünnow attend

1915   *Publications of the Observatory of the University of Michigan,*
         Vol. I, Part II, is published
         Huddy Hill is considered again as a possible relocation sight
         for the Observatory

1916   *Publications of the Observatory of the University of Michigan,*
         Vol. II, is published
         Eric Brünnow dies of infantile paralysis

1917   Course in navigation is introduced for the benefit of
         U.S. Naval Reserve units
         Rudolph Brünnow dies at Bar Harbor, Maine

1922   37½-inch reflecting telescope overhauled; driving clock improved

1923   Students' Observatory dismantled to make way for Couzens Hall
         Publications of the Observatory of the University of Michigan,
         Vol. III, is published

1924   Two-prism spectrograph designed by Curtiss is constructed in
         the Observatory Shop

Hartman spectrocomparator is purchased
Power Plant construction on the Cat-Hole site encroaches on
    the Observatory with smoke and limited sight distance
Land near Portage Lake identified as possible relocation site
Hussey offers a popular extension course in astronomy in Detroit
Simpson Building to be constructed on Observatory land
    to the east

1925   Hussey and others attempt to observe a solar eclipse from a
            balloon near Geneva, New York, but clouds and wind
            prevent it; Rufus is stationed at Bad Axe, where it is
            also cloudy
January   Lamont 27-inch refractor completed
July   Lamont refractor set up on Observatory lawn for testing
            prior to shipment to South Africa

1926   Ralph H. Curtiss becomes Acting Director
            Harrington dies
August   Lamont 27-inch refractor is shipped to South Africa
October   During travel to Bloemfontain, South Africa, undertaken
            despite being ill with pleurisy, Hussey dies in London
November Richard A. Rossiter takes over in South Africa

1927   Ralph H. Curtiss becomes Director
            Angell Hall Observatory opened for student use; two 20-foot
            Fecker domes house 10-inch Warner & Swasey refractor
            (1926) and 15-inch Fecker reflector (1929)
            Hazel "Doc" Losh (Ph.D. '24) joins the Astronomy Department
            as a Research Assistant in Astrophysics

1928   Option secured on 200 acres of land near Portage Lake;
            availability of land presently occupied by the Observatory
            was offered to make the deal more attractive
            Lamont-Hussey Observatory is dedicated

1929   Ralph H. Curtiss dies
            W. Carl Rufus becomes Acting Director
            Watson's claim that he discovered the planet Vulcan is finally
            dismissed

1930   Heber D. Curtis becomes Director
            Howard sidereal clock and Hale spectrohelioscope purchased
            for Angell Hall Observatory

Moll microphotometer from Kipp and Zonen purchased for the
Observatory
85½-inch Pyrex mirror produced for a new telescope, but found
faulty
97½-inch Pyrex mirror produced for a new telescope that was
never completed; mirror placed in storage near Observatory
Asaph Hall dies
McMath-Hulbert Observatory opens at Lake Angeles

1931   Board of Regents officially changes the name of the Detroit
Observatory to the inclusive "Observatories of the
University of Michigan"
Curtis improves slow-motion guiding of the 37½-inch
reflecting telescope

1932   Curtis observes solar eclipse in Fryeburg, Maine
Publications of the Observatory of the University of Michigan,
Vol. IV, is published

1934   37½-inch mirror is aluminized

1935   *Publications of the Observatory of the University of Michigan,*
Vol. V, is published

1937   *Publications of the Observatory of the University of Michigan,*
Vol. VI, is published

1939   *Publications of the Observatory of the University of Michigan,*
Vol. VII, is published
Henry Fitz, Jr. dies

1941   W. Carl Rufus becomes Director

1942   Heber D. Curtis dies

1945   Alan D. Maxwell becomes Acting Director
Land purchased at Peach Mountain

1946   Leo Goldberg becomes Director

1960   Freeman D. Miller becomes Acting Director

<u>1960s</u>	Observatory largely abandoned; students and others use it for unauthorized activities
<u>1961</u>	Orren C. Mohler becomes Director
<u>1963</u>	Astronomy moves into the new Dennison Building on central campus; Observatory used as Astronomy Library
<u>1966</u>	Plan developed to move the Observatory 20 feet to the south to expand the intersection in front of Old Main Hospital (never acted on)
<u>1970</u>	W. Albert Hiltner becomes Director Bookcases are removed from the west wing Asphalt roll roofing over fiberglass insulation is added to the Observatory
<u>1970s</u>	University proposes to raze the buildings; friends of the Observatory appeal to save the original building, with success
<u>1973</u>	Observatory is listed in the National Register of Historic Places
<u>1976</u>	1908 addition is razed. Mountings from the 7½-inch reflecting telescope given to the Lake Erie Astronomical Project; mirror and driving clock placed in storage
<u>1979</u>	Chronograph by Fauth & Co. sent out for restoration
<u>1980s</u>	Building used as a warehouse for historical scientific and medical apparatus, and as the headquarters for the Collegiate Institute for Values and Science Second floor modified for use as office space
<u>1989</u>	Building receives repairs and painting in preparation for the annual meeting of the American Astronomical Society held in Ann Arbor Meridian circle is cleaned and polished by Orren Mohler
<u>1991</u>	W. Albert Hiltner dies; memorial plaque and garden installed on the grounds of the Detroit Observatory

1994
March   Oversight of the Detroit Observatory transferred to the Office
               of the Vice President for Research in response to its proposal
               to restore the building
               Detroit Observatory Advisory Group appointed by
               Vice President for Research, Homer A. Neal; Whitesell
               appointed as Chair

1994-95   Scientific and medical apparatus stored in the Observatory is
               relocated, building is stabilized and made usable for
               fundraising activities

1996   Restoration planning process launched with historic
               preservation architect
               Preservation Study is completed; copy placed at Bentley
               Historical Library
               Fundraising continues
               President Duderstadt, Vice President Neal, Associate Vice
               President Neidhardt, and family members had an evening
               of stargazing at the Detroit Observatory
               Regent Emeritus Alfred Connable was fêted at the Observatory
               for designating a bequest; present were President
               Duderstadt, several former presidents, Regents, and others
               important in the University's history
               Construction of the Magellan Project observatory in Chile

1997-98   Restoration construction process
               Every step of the restoration work is photographically
               documented
               Architectural drawings, daily field logs, etc., placed at
               Bentley Historical Library
               Fitz refractor is restored; consultant's journal placed at
               Bentley Historical Library
               Scale model of original Fitz with its wooden tube is constructed
               Chief Historian of the National Park Service visits the Detroit
               Observatory
               Vice President Neidhardt tours the restoration in progress
               President Bollinger tours the restoration in progress
               Whitesell appointed as Director and Curator of the Observatory
               Provost Cantor tours the newly restored building
               Publication of *A Creation of His Own: Tappan's Detroit Observatory*

planned   Building dedication
                 Creation of museum exhibits and interior furnishing
                 Program planning and implementation
                 Installation of historical plaque by the University's History and
                     Traditions Committee

planned   Celebration of the Detroit Observatory's 150th anniversary
for 2004   Rare transit of Venus will take place; the last transit was
                     observed in 1882

Notes

Introduction

1. Michigan Senate Document, No. 28, February 26, 1850. The titles chancellor and president were used interchangeably when Tappan was being recruited, but a Detroit newspaper reporter named W. F. Storey later accused Tappan of pretension because he used the title Chancellor.

2. Tappan, Henry P., *Review by Rev. Dr. H. P. Tappan of His Connection with the University of Michigan*, Free Press Steam Power Press Printing House, Detroit, 1864, p. 17.

3. The Detroit Observatory and its telescopes cost about $22,000 (approximately $330,000 in 1998 dollars), compared to the Dudley Observatory in Albany, NY for which in excess of $160,000 was raised.

4. "First" refers to quality and ranking rather than to the order of creation.

5. University of Michigan, *Regents Proceedings, 1837-1864*, October 1854, p. 600.

6. Tappan, Henry P., *Review by Rev. Dr. H. P. Tappan*, 1864, p. 17.

Chapter 1. Henry Philip Tappan

7. University of Michigan, *Regents Proceedings, 1837-1864*, p. 512-521.

8. University of Michigan, Bentley Historical Library, Charles H. Palmer family papers, Box 1, letter from Henry P. Tappan to Charles H. Palmer, March 15, 1852.

9. For a complete biography of Tappan's life, see Perry, Charles M., *Henry Philip Tappan: Philosopher and University President*, University of Michigan Press, Ann Arbor, 1933.

10. Union College, *Union Worthies*, No. 22, "Henry Philip Tappan, Class of 1825," by Kenneth T. Doran and Harlan H. Hatcher, Schenectady, New York, 1967, p. 5-11.

11. Hislop, Codman, *Eliphalet Nott*, Wesleyan University Press, 1971, p. 231.

12. Ibid., p. 229.

13. Ibid., p. 109.

14. Ibid., p. 110.

15. Ibid., p. 111. For a fascinating account of early life on the Michigan frontier, read Cooper's *Oak Openings*.

16. Ibid., p. 110.

17. This photograph is of the solar eclipse of 1869, taken at Mount Pleasant, Iowa by Professor Henry Morton and a group of skilled photographers. See Chapter 11

for a complete account of the role the University of Michigan played in observing the eclipse. See also Figures 11.3 and 11.4.

18. Hislop, Codman, *Eliphalet Nott,* Wesleyan University Press, 1971, p. 212.
19. Ibid., p. 213.
20. University of Michigan, *Michigan Alumnus,* No. 6, May 1902, p. 370. Tappan's wedding date and family birth and death dates were recorded in the Tappan family *Bible,* which was donated to the University in 1902 by Gabriel Campbell, Class of 1865, professor of philosophy at Dartmouth College. The 1,214-page book was printed for the American Bible Society in 1834 as a third edition. The current location of this *Bible* is not known.
21. Union College, *Union Worthies,* No. 22, "Henry Philip Tappan, Class of 1825," by Kenneth T. Doran and Harlan H. Hatcher, Schenectady, New York, 1967, p. 5.
22. Tappan's college roommate in Room 12 of North College at Union College was from Pittsfield, Massachusetts, a fact which may help to explain Tappan's ministerial appointment in this city.
23. Jones, Theodore. F. (ed.), *New York University, 1832:1932,* New York University Press, 1933, p. 49-50.
24. Perry, Charles M., *Henry Philip Tappan,* 1933, p. 85; Williams, Edwin, *New York As It Is,* J. Disturnell, New York, 1834, p. 47. For an explication of Tappan's views on educating women, see Perry, Charles M., *Henry Philip Tappan,* 1933, p. 362, and Frieze, Henry S., *A Memorial Discourse on the Life and Services of Henry Philip Tappan, D.D., LL.D.,* University of Michigan, 1882, p. 23.
25. Frieze, Henry S., *A Memorial Discourse,* 1882, p. 20.
26. Runes, Dagobert D., *Pictorial History of Philosophy,* Philosophical Library, New York, 1959, p. 204.
27. The Nott Memorial stands today, having undergone an extensive restoration in 1994-95.
28. University of Michigan, *Michigan Alumnus,* No. 6, May 1902, p. 370.
29. Tappan, H. P., *A Discourse delivered by Henry P. Tappan, D.D., at Ann Arbor, Mich., on the Occasion of his Inauguration as Chancellor of the University of Michigan, December 21, 1852,* Detroit, 1852, p. 4.
30. Ibid., p. 5-6.
31. Ibid., p. 18-19.
32. Farrand, Elizabeth M., *History of the University of Michigan,* Ann Arbor, 1885, p. 16.
33. University of Michigan, *Michigan Alumnus,* August 1904, p. 580. This is the reminiscence of Colonel Isaac H. Elliott, Class of 1861.

Chapter 2. The Emergence of American Astronomical Science

34. Ambrose, Stephen E., *Undaunted Courage,* Simon and Schuster, New York, 1996, p. 119-120.
35. *Nature,* "Observatories in the United States," July 9, 1874, p. 186.
36. *Centenary of the Cincinnati Observatory,* November 5, 1943, Cincinnati, 1944, p. 27.
37. Figure 3.4 is a painting of Lewis Cass, shown with a telescope, books, and a globe—images used to depict his elite position in society.

38. Jones, Bessie Zaban, *The Harvard College Observatory: The First Four Directorships, 1839-1919*, Belknap Press of Harvard University Press, 1971, p. 36.
39. *Centenary of the Cincinnati Observatory*, November 5, 1943, Cincinnati, 1944, p. 20.
40. Watson, James C., *A Popular Treatise on Comets*, James Challen & Son, Philadelphia, 1861, p. vii.
41. Bruce, Robert V., *Launching of Modern American Science, 1846-1876*, Alfred A. Knopf, New York, 1987, p. 101.
42. Rudolph, Frederick, *The American College and University*, Vintage Books, New York, 1962, p. 225-226.
43. Mehlin, Theodore G., "Williams College Renovates Hopkins Observatory," a photocopied article given to the author by Professor Jay Pasachoff at Williams College. The original telescope at Hopkins Observatory was replaced in 1852 with a 7-inch Alvan Clark objective with a Jonas Phelps mount, which is presently in use for viewing nights. This telescope may have been the first commercially-made by Clark.
44. This octagonal observatory, completed in 1848, was designed by Henry A. Sykes at the direction of President Hitchcock. No doubt, Hitchcock was influenced by Orson S. Fowler, Amherst Class of 1834, who invented and published the definitive text on octagonal buildings, *A Home For All*, in 1850.
45. The Hamilton College Observatory was endowed in 1866 through the liberality of E. C. Litchfield of Brooklyn, New York, and renamed the Litchfield Observatory in his honor. Litchfield was an alumnus of Hamilton College, and a donor to the Detroit Observatory. It was later proposed that the Litchfield Observatory be moved to Prospect Park where the Litchfield mansion [see Figure 9.4], designed by Alexander Jackson Davis (architect of the original plan for the University of Michigan campus), was located. [See Chapter 3]
46. See *Harper's New Monthly Magazine*, Elias Loomis, "Astronomical Observatories in the United States," Part I, Vol. 13, June 1856, p. 25-52; "Observatories in the United States," Part II, Vol. 48, March 1874, p. 518-539, for an extensive description of many early observatories.
47. *Harper's New Monthly Magazine*, "The Coast Survey," Vol. 58, March 1879, p. 520.
48. For a fascinating account of the longitude problem, that is easily understood by the layperson, see Dava Sobel's *Longitude*, Walker & Co., New York, 1995; and *The Quest for Longitude*, William J. H. Andrewes (ed.), Harvard University, 1998.
49. Farmer, Silas, *The Michigan Book: A State Cyclopedia*, Silas Farmer & Co., Detroit, 1901, p. 302.
50. William A. Burt also invented Burt's Typographer, the first typewriter, in 1828.
51. Burt, John S., *They Left Their Mark: William Austin Burt and His Sons, Surveyors of the Public Domain*, 1985, p. 70.
52. *Harper's New Monthly Magazine*, "The Electric Time Service," Vol. 56, October 1878, p. 665.
53. Ibid., "Astronomical Observatories in the United States," Part I, Vol. 13, June 1856, p. 52.
54. Ibid., "Can We Foretell the Weather?," Vol. 39, August 1869, p. 422.
55. This windvane, handmade of forged metal with wooden blades and ball, is still extant.
56. These ledgers are still extant.

Chapter 3. Tappan's Observatory: The Detroit Donors

57.   Tappan, Henry P., *A Discourse, delivered by Henry P. Tappan*, 1852, p. 17-18.
58.   Ibid., p. 32.
59.   University of Michigan, *Encyclopedic Survey*, Vol. II, p. 465 (original source is *Report of the Superintendent of Public Instruction of the State of Michigan*, 1849).
60.   Olmsted, Dennison, *A compendium of astronomy; containing the elements of the science, familiarly explained . . . , with the latest discoveries*, Collins, Keese & Co., New York, 1839.
61.   Tappan, Henry P., *A Discourse, delivered by Henry P. Tappan*, 1852, p. 16.
62.   Ibid., p. 51.
63.   Stephenson, O. W., *Ann Arbor: The First Hundred Years*, Ann Arbor, 1927, p. 331; Farmer, Silas, *The Michigan Book: A State Cyclopedia*, Detroit, 1901, p. 897.
64.   Farmer, Silas, *The Michigan Book: A State Cyclopedia*, Detroit, 1901, p. 212.
65.   J. W. Brooks was a donor to the Detroit Observatory.
66.   Thompson, Robert L., *Wiring a Continent: The History of The Telegraph Industry in the United States, 1832-1866*, Princeton University Press, 1947, p. 206.
67.   Elon Farnsworth and James F. Joy were donors to the Detroit Observatory.
68.   The Michigan Exchange, which opened in 1835, was located at the southeast corner of Jefferson Avenue and Shelby Street.
69.   University of Michigan, *Regents Proceedings*, 1837-1864, October 1854, p. 576-577.
70.   University of Michigan, *University Magazine*, July 1869, p. 389.
71.   *Harper's New Monthly Magazine*, "Application of Electro-magnetic Power to Railway Transit," Vol. 3, November 1851, p. 787.
72.   University of Michigan, Bentley Historical Library, Palmer Family papers, Box 1, letter from Henry P. Tappan to Charles H. Palmer dated November 5, 1853.
73.   University of Michigan, Bentley Historical Library, Andrew D. White papers, letter dated July 4, 1861 from Lizzie to Aimèe.
74.   Ibid., letter from H. P. Tappan to C. H. Palmer from Stuttgart dated January 10, 1868; letter from HPT to CHP from Nice dated November 23, 1869; letter from HPT to CHP from [Bad] Homburg dated March 28, 1872.
75.   For further reading see *Michigan History*, "The Detroit Young Men's Society," June 1959, p. 197-211.
76.   Moore, Charles, *History of Michigan*, Lewis Publishing Company, Chicago, 1915, Vol. I, p. 442.
77.   Cooley, Thomas McIntyre, *Michigan: A History of Governments*, Houghton, Mifflin and Company, New York, 1897, p. 239.
78.   Descendants of Frederick and Christian Buhl made contributions to the University of Michigan in the 20th century, including the Buhl Research Center for Human Genetics and the Lydia Mendelssohn Theatre, named in memory of Lydia Mendelssohn Buhl by her son.
79.   See footnotes 45, 242, and 259.
80.   Brush's son, Charles Francis, was an electrician who also constructed telescopes.
81.   Bela Hubbard's fascinating writings include *Lake Superior Journal: Bela Hubbard's Account of the 1840 Houghton Expedition* (1983); *Memorials of a Half-Century* (1887); and *Reports of William A. Burt and Bela Hubbard* (1846).

82. George Duffield donated to the University of Michigan Library on February 2, 1884 his volumes of the *Nautical Almanac and Astronomical Ephemeris* (see book plate). Richard Harrison Bull (see Chapter 8), who designed the Detroit Observatory, was an elder of the First Presbyterian Church of New York, and in 1891, Bull nominated the Reverend Howard Duffield of Detroit as pastor. Dr. Duffield accepted. Other evidence that George Duffield was interested in astronomy appears in Warner and Ariail's book *Alvan Clark & Sons: Artists in Optics*, 2nd ed. (p. 149); although the dates reported are off, George Duffield reportedly purchased in 1907 (he died 1888) a 6-inch Clark telescope from the Raisin River Seminary at Adrian, MI, where Duffield spent four years in the ministry (1861-1865).

83. Biographical information on the donors is from the *National Cyclopaedia of American Biography;* Moore's *History of Michigan,* 1915; *Representative Men of Michigan,* 1878; *American biographical history of eminent and self-made men,* Michigan Volume, 1878; and numerous other sources.

84. University of Michigan, Bentley Historical Library, Alexander C. McGraw papers, letters dated December 15, 1856 and December 9, 1857.

85. University of Michigan, Bentley Historical Library, Lucy Chapin papers, folder 3, scrapbook, undated, unidentified newspaper article.

86. Dudley Observatory Archives, James H. Armsby Collection, Box 1, folder 23, letter from O. M. Mitchel to Executive Committee, Albany Observatory dated January 13, 1853.

87. Harvard University Archives, William C. Bond papers, letter from W. C. Bond to F. Brünnow dated December 22, 1858.

88. Burt, John S., *They Left Their Mark,* 1985, p. 51.

89. For a detailed explanation of the calculation of time using a small transit instrument, see *Sidereal Messenger,* Vol. III, 1884, p. 209-13.

90. Harvard University Archives, William C. Bond papers, letter from Brünnow to W. C. Bond dated December 30, 1858.

91. University of Michigan, Bentley Historical Library, Alexander Winchell papers, scrapbook, p. 148, unidentified 1863 newspaper account from a Detroit paper. Smith's jewelry store was at 55 Woodward Avenue in Detroit.

92. Dropping the timeball at noon was the standard in America; 1:00 p.m. was the standard in Europe.

93. U.S. Naval Observatory, *Sky With Ocean Joined,* "Naval Observatory Time Dissemination Before the Wireless," by Ian R. Bartky, 1983, p. 7.

94. The Railroad Depot was located where the Gandy Dancer restaurant is located today. The stone structure currently on this site was constructed in 1887. The original depot, built in 1839, was relocated to the southeast corner of Beakes St. and North Fifth Ave. to make way for the new station, and it is still extant.

95. Harvard University Archives, W. C. Bond papers, letter from F. Brünnow, Ann Arbor, to G. P. Bond dated October 1, 1860.

96. Tappan, Henry P., *Review by Rev. Dr. H. P. Tappan,* 1864, p. 31.

97. University of Michigan, *Regents Proceedings,* 1837-1864, p. 918.

98. Harvard University Archives, Bond Papers, letter dated October 1, 1860 from F. Brünnow, Ann Arbor, to G. P. Bond.

99. The Detroit Observatory has in its collection actual sheets of the paper that came off the chronograph drum, dating back to October 9, 1860, shortly after the chronograph was obtained. The sheets contain markings noting the position of stars, and notations made by Brünnow and other astronomers. [see Figure 3.13] The sheets were being discarded during file purging in the Atmospheric, Oceanic and Space Sciences department around 1980 when a faculty member interested in history retrieved them and kept them in his office, returning them to the Observatory in 1996. Many thanks go to Dennis Baker for saving these treasures. Thanks also go to Paul Gross for saving important weather data.

100. For an explanation of Cyrus W. Field's efforts to install a telegraph cable from Newfoundland across the Atlantic Ocean, see his biographical sketch, *National Cyclopaedia of American Biography*, New York, 1898, p. 451-453.

101. Farmer, Silas, *History of Detroit and Michigan*, Silas Farmer & Co., Detroit, 1884, p. 884-885.

102. *Astronomical Notices*, No. 27, October 8, 1861, Ann Arbor, p. 17-18.

103. University of Michigan, *Encyclopedic Survey*, Vol. II, p. 447.

104. University of Michigan, *Regents Proceedings*, 1864-1869, June 29, 1864, p. 46.

105. University of Michigan, "Report of the Curator of the University Museum to the Board of Regents, July 1, 1903 to July 1, 1904," Ann Arbor, 1905, p. 12.

Chapter 4. Pistor & Martins Meridian Circle

106. University of Michigan, *Regents Proceedings*, 1837-1864, p. 630.

107. Sturges, Mrs. Jonathan, *Reminiscences of a Long Life*, New York, 1894 (1996 reprint of an autobiography published privately in 1894. Carol Mull provided this family history). Tappan's role in escorting the children to Europe is described in the *Detroit Free Press*, December 28, 1853, as that of tutor.

108. *The Michigan Argus*, Wednesday, December 14, 1853, reprinted from the Report to the Board of Regents of the University of Michigan, Nov. 15, 1853, by Henry Philip Tappan.

109. *Albany Argus*, "The Detroit Observatory of the University," 1858, month and date unknown; document from the C. H. F. Peters papers, Hamilton College Library, Clinton, New York.

110. Sidereal time varies on a daily basis from standard time by about four minutes. The sidereal day is based on the interval between two successive transits of the March equinox over the upper meridian of a place, which takes 23 hours, 56 minutes, and 4.09 seconds.

111. Andrews, A. D., *Irish Astronomical Journal*, "Cyclopaedia of Telescope Makers," Parts 3 and 4, Vols. 21 and 22, 1994 and 1995, p. 213 and 57, respectively.

112. For more details on the functioning of a meridian circle telescope and its appliances, see Brünnow's *Astronomical Notices*, "On the Determination of the Flexure of Astronomical Instruments," No. 26, August 14, 1861; Wayman, Patrick A., *Dunsink Observatory, 1785-1985: A Bicentennial History*, Royal Dublin Society, 1987, appendix 17: "Brünnow's Description of the Transit Circle," p. 314-315; Ambronn, L., *Handbuch der Astronomischen Instrumentenkunde*, Berlin, 1899; Greene, Dascom, *Spherical and Practical Astronomy*, Boston, 1891; Kuiper

and Middlehurst (eds.), *Telescopes*, Chicago, 1960, Chapter 6, "Transit Instruments," p. 80-87.

113. The original weather instruments are no longer extant, though numerous weather instruments from later periods are held in the Detroit Observatory's collection of scientific instruments. An Ann Arbor man named L. Woodruff received a patent in 1860 for his barometer. It is not known whether his instrument was used or tested at the Detroit Observatory. See W. E. Knowles Middleton, *The History of the Barometer*, The Johns Hopkins Press, 1964. A Hough's barograph was installed in the Observatory in the late 1860s, an instrument designed by G. W. Hough of the Dudley Observatory. Brünnow very likely brought this instrument back with him from Albany. A photograph of the barograph is in the archives of the Bentley Historical Library.

114. *Albany Argus*, "The Detroit Observatory of the University," 1858, month and date unknown; document from the C. H. F. Peters papers, Hamilton College, Clinton, New York; *Annals of the Dudley Observatory*, Vol. I, Albany, 1866, p. 21.

115. *Sidereal Messenger*, Vol. III, 1884, p. 206.

Chapter 5. Henry Fitz, Jr.: American Telescope Maker

116. Henry Fitz's telescope shop was donated by the Fitz family to the Smithsonian Institution and was reconstructed at the U.S. National Museum; the exhibit is currently in storage, as is the 12-inch Fitz refractor from Vassar College.

117. Abrahams, Peter, "Henry Fitz: A Preeminent 19th Century American Telescope Maker," *Journal of the Antique Telescope Society*, Summer 1994, p. 6-10.

118. Multhauf, Robert P., *Holcomb, Fitz, and Peate: Three 19th Century American Telescope Makers*, United States National Bulletin 228, Smithsonian Institution, Washington, D.C., 1962, p. 165.

119. A speculum is a reflector (mirror) in an optical instrument, including telescopes and cameras.

120. Newhall, Beaumont, *The Daguerreotype in America*, Duell, Sloan & Pearce, 1961, p. 23.

121. Rhinhart, Floyd and Marion, *The American Daguerreotype*, University of Georgia Press, Athens, 1981, p. 30.

122. See the Dorothy Draper daguerreotype in Scharf, Aaron, *An Album of Pictures and Words*, Harry N. Abrams, Inc., New York, 1975, illustration 2.3.

123. Draper, John William, *Scientific Memoirs being Experimental Contributions to A Knowledge of Radiant Energy*, Arnor Press, New York, 1973, p. 215.

124. See the Susan Fitz daguerreotype in Scharf, Aaron, *An Album of Pictures and Words*, Harry N. Abrams, Inc., New York 1975, illustration 2.4.

125. Multhauf, Robert P., "Holcomb, Fitz, and Peate: Three 19th Century American Telescope Makers," *United States National Bulletin 228*, Smithsonian Institution, Washington, D.C., 1962, p. 168.

126. To some, the handsome Henry Fitz resembles Marlon Brando.

127. University of Michigan, Bentley Historical Library, Henry Philip Tappan papers, folder 2, February 1868. The Livingstons financed the experiments of Robert Fulton that led to the launching of the Clermont, the first commercially

successful American steamboat. The Livingstons thereafter held a monopoly
on steamboat operations in New York waters.

128. The University of Michigan purchased twenty feet of Atlantic cable for $12.63
from Tiffany & Co. in September 1858 (see *Regents Proceedings*, 1837-1864,
University of Michigan, p. 758). How this cable was used is unknown.

129. Rinhart, Floyd and Marion, *The American Daguerreotype*, University of Georgia
Press, Athens, 1981, p. 390.

130. Multhauf, Robert P., "Holcomb, Fitz, and Peate: Three 19th Century American
Telescope Makers," *United States National Bulletin 228*, Smithsonian Institution,
Washington, D.C., 1962, p. 168.

131. Sperling, Norman, "Fair Play for Fitz: Henry Fitz Introduces the All-American
Telescope," *Rittenhouse*, Journal of the American Scientific Instrument Enterprise,
Vol. 3, No. 2, Issue 10, February 1989, p. 51.

132. Loomis, Elias, *Recent Progress of Astronomy*, Harper & Brothers, New York, 1856;
Gilliss, James M., *U.S. Naval Astronomical Expedition to the Southern Hemisphere*,
1849-50-51-52, Vol. III, "Observations to Determine the Solar Parallax,"
Washington, 1856, p. xxv, xlvi.

133. Mattison, Hiram, *A High School Astronomy*, New York, 1857, p. 225. Note that this
book was subsequently titled *Geography of the Heavens*, and authorship was
attributed to Elijah Burritt, though the text remained largely the same. Editions
vary in content.

134. Gilliss, James M., in *U.S. Naval Astronomical Expedition to the Southern Hemisphere*,
1849-50-51-52, Vol. III, "Observations to Determine the Solar Parallax,"
Washington, 1856, p. xxv, xlvi.

135. *Scientific American*, "The Largest Telescope in the Country," Vol. 4, No. 14 new
series, April 6, 1861, p. 216.

136. Warner, Deborah Jean, "Elisa Robinson and Henry Fitz," *Rittenhouse*, Journal of
the American Scientific Instrument Enterprise, Vol. 1, No. 1, Issue 1, November
1986, p. 19-20.

137. Sperling, Norman, "Fair Play for Fitz," p. 50.

138. Zahner, Donald D., *Popular Astronomy*, "Henry Fitz, Telescope Maker,"
Vol. LVIII, No. 529, August/September 1964, p. 6.

139. Sperling, Norman, "Fair Play for Fitz," p. 50.

140. University of Michigan, Bentley Historical Library, James C. Watson papers,
jotting book, pages numbered 161-163, transcribed from a letter from Henry Fitz
to James C. Watson dated March 30, 1856.

141. Rinhart, Floyd and Marion, *The American Daguerreotype*, University of Georgia
Press, Athens, 1981, p. 431.

142. Trombino, Don and Pazmino, John, "Henry Fitz: Locksmith, Photographer, and
Telescope Maker," *Skylines*, 1977, p. 5.

143. Chamberlain, Joshua L. (ed.), *Universities and Their Sons*, R. Herndon Co.,
Boston, 1898, p. 27.

144. This essay of 121 pages was among the Tappan papers Rudolph Brünnow
transmitted to the University of Michigan Library in March 1910 at the request
of Charles M. Perry, who wrote Tappan's biography; it is now at the Bentley
Historical Library.

145. *Albany Evening Journal*, "Astronomers and Astronomical Instruments," August 30, 1856, p. 2, column 4.
146. Newington-Cropsey Foundation, letter from Henry P. Tappan to Maria Cropsey, July 28, 1856. It is interesting to note in Tappan's letter that he reported he was suffering from writer's block due to the dry, intense July heat of Michigan.
147. Smithsonian Institution Archives, National Museum of American History, letter from Charles W. Hackley to Henry Fitz, Jr. dated only August. Mention is made of a scientific meeting in Albany, which was likely the AAAS meeting of August 28, 1856, coincident with the dedication of the Dudley Observatory.
148. *Albany Argus*, "The Detroit Observatory of the University," 1858, month and date unknown; document from the Hamilton College Archives, C. H. F. Peters Collection.
149. *Sidereal Messenger*, Mitchel, O. M. (ed.), Cincinnati, August 1846, Vol. 1, No. 2, p. 13.
150. Sperling, Norman, "Fair Play for Fitz," p. 50.
151. *Albany Argus*, "The Detroit Observatory of the University," 1858, month and date unknown; document from the Hamilton College Archives, C. H. F. Peters Collection.
152. Mattison, Hiram, *A High School Astronomy*, New York, 1857, p. 225.
153. University of Michigan, *Regents Proceedings, 1837-1864*, p. 647, 705.
154. Ibid., Bentley Historical Library, Alexander Winchell papers, Winchell scrapbook, p. 103, from a newspaper article written for the *Chicago Tribune* in the winter of 1861 regarding the meeting of the Chicago Academy of Sciences in Ann Arbor, which included a tour of the Detroit Observatory.
155. This was one of Clark's first commercially-made telescopes.
156. Warner, Deborah Jean and Ariail, Robert B., *Alvan Clark & Sons: Artists in Optics*, second edition, Willman-Bell, Inc., Richmond, VA, 1995, p., 114, 181. The Clarks were competitors of Fitz, eventually taking control of the telescope market after Fitz's death in 1863. Note also that Charles W. Hackley, Tappan's former colleague at NYU and friend of Fitz, had been President of Jefferson College until 1857.
157. The Detroit Observatory's museum has in its collection two compasses made by W. & L. E. Gurley of Troy, NY.
158. Smart, Charles E., *The Makers of Surveying Instruments in America Since 1700*, Regal Art Press, Troy, NY, 1962, p. 116-119.
159. Hall, Asaph, *Determination of the Aberration Constant*, Lansing, MI, 1904, p. 38.
160. Doug Covert performed the calculations in 1997.
161. Not trusting the dome's shutter opening to support the weight of the telescope tube in 1997 when the Fitz refractor was fully restored, scaffolding was erected in the dome to support an I-beam above the telescope on which ran a block and tackle. The telescope tube was lowered and re-mounted using a chain, spreader bar, and soft straps.
162. University of Michigan, *Regents Proceedings, 1837-1864*, October 1856, p. 678.
163. University of Michigan, *Encyclopedic Survey*, Vol. II, p. 470.

Chapter 6. The German Connection: Recruitment of Franz Brünnow

164. University of Michigan, *Michigan Alumnus*, Vol. 9, 1903, p. 419.
165. Tappan, Henry P., *Review by Rev. Dr. H. P. Tappan*, 1864, p. 9.
166. For a fascinating account of the early history, and the controversy surrounding, the establishment of the Dudley Observatory, see Mary Ann James' *Elites in Conflict*, Rutgers University Press, 1987; or, read the original documents (see Bibliography under Dudley Observatory).
167. Humboldt's brother, Wilhelm von Humboldt, was the Prussian minister of education and one of the founders of the University of Berlin.
168. University of Michigan, Bentley Historical Library, Alexander Winchell papers, Winchell scrapbook, p. 2, news clipping from the *Detroit Advertiser*, undated. This letter, translated from the German, was written in Berlin on May 4, 1857, at the time Brünnow was visiting Europe, on leave for eight months following his wedding. Humboldt may have been moved to write it based on accounts he heard from Brünnow of mistreatment in the press by University Regent, Levi Bishop.
169. University of Michigan, Bentley Historical Library, William D. Anderson papers, letter dated August 21, 1854 written by Anderson from Ann Arbor.
170. Frieze, Henry S., "Reminiscence of the Times of President Tappan," *Palladium*, Ann Arbor, 1888, p. 113-128.
171. University of Michigan, Bentley Historical Library, Henry S. Frieze papers, letter from H. S. Frieze to his wife, Anna, from Berlin dated November 15, 1855.
172. Ibid., letter from H. S. Frieze to his wife, Anna, from Berlin dated February 21, 1856.
173. *Sidereal Messenger*, Vol. III, 1884, p. 213. Through a powerful telescope, some stars may appear double because of the optical effect of their position relative to the observer. When stars are placed nearly in the same line of vision, they will appear contiguous, though they may actually be a great distance apart.
174. *Albany Evening Journal*, "Astronomers and Astronomical Instruments," August 30, 1856, p. 2, column 4.
175. Kestenbaum, Justin L. (ed.), *The Making of Michigan, 1820-1860*, Wayne State University Press, Detroit, 1990, p. 264.
176. University of Michigan, Bentley Historical Library, Henry S. Frieze papers, letter from J. R. Boise to H. S. Frieze from Ann Arbor dated January 2, 1856.
177. Newington-Cropsey Foundation, letter from Henry P. Tappan to Jasper F. Cropsey from New York dated May 12, 1857.
178. Perry, Charles M., *Henry Philip Tappan*, 1933, p. 162.
179. Copies of these works were presented by Rebecca Brünnow to the University of Michigan Libraries, and bear her inscription on the inside covers. These volumes and other works by Tappan, Brünnow, Watson, etc. are now on permanent loan to the Detroit Observatory.
180. University of Michigan, Special Collections Library, Edward Payson Evans papers, letter from Rudolph Brünnow to Edward Payson Evans from Vevey, Switzerland dated May 6, 1893.
181. Vesta was the fourth asteroid ever discovered, in 1807 by Olbers of Bremen, Germany.

182. Victoria was the twelfth asteroid ever discovered, by Hind of London in 1850.
183. Peter Cooper was an American inventor, industrialist, and philanthropist known for his involvement in railroad and telegraph operations, making his early fortune in the glue business and later in iron works. One of the earliest locomotives built in the U.S., the *Tom Thumb*, was built by Cooper.
184. Samuel Bulkley Ruggles had been a prominent lawyer in New York, but gave up his practice to devote himself entirely to public affairs. He was responsible for laying out Gramercy Park on land he donated, and Union Square; was involved in the Croton Reservoir project; helped to organize the Erie and Union Pacific Railroads; and was a trustee of Columbia College.
185. University of Michigan, Bentley Historical Library, Henry Philip Tappan papers, folder 1, letter from H. P. Tappan to S. B. Ruggles from Ann Arbor dated January 29, 1856. See also letters from HPT to SBR dated Feb. 19, Feb. 20 and April 28, 1856.
186. This reference is most likely to William Backhouse Astor (1792-1875) of New York, son of John Jacob Astor (1763-1848); W. B. Astor was an American financier, referred to as the "landlord of New York."
187. University of Michigan, Bentley Historical Library, Henry P. Tappan papers, folder 1, letter from H. P. Tappan to Hon. H. Barney from Ann Arbor dated June 13, 1863.
188. For a thorough account of Tappan's removal from office, see Charles Perry's *Henry Philip Tappan*, 1933; and, Tappan, Henry P., *Review by Rev. Dr. H. P. Tappan*, 1864.
189. Harvard University Archives, W. C. Bond papers, letter from J. C. Watson to G. P. Bond from Ann Arbor dated August 29, 1863.
190. Ibid., W. C. Bond papers, G. P. Bond's letterpress copy of a letter he wrote to J. C. Watson dated October 3, 1863, Cambridge, MA.
191. Royal Astronomical Society, *Monthly Notices*, Vol. LII, 1892, p. 232.
192. University of Michigan, *Michigan Alumnus*, August 1904, p. 584.
193. Perry, Charles M., *Henry Philip Tappan*, 1933, p. 382-383. Letter from Tappan to Mr. Murphy from Dunsink Observatory dated August 17, 1866.
194. Royal Astronomical Society, *Monthly Notices*, obituary, Vol. LII, London, 1892, p. 232.
195. Perry, Charles M., *Henry Philip Tappan*, 1933, p. 409.
196. Eric Brünnow, who was attending Princeton University, died two years later on October 16, 1916 of "infantile paralysis," as reported in the *Michigan Alumnus*, November 1916, p. 64.
197. University of Michigan, *Michigan Alumnus*, August 1914, p. 577.
198. Wayman, Patrick A., *Dunsink Observatory, 1785-1985: A Bicentennial History*, Royal Dublin Society, 1987, Appendix 16. Rudolph Brünnow was age fourteen in 1872.
199. Rudolph inherited the Tappan estate, which would have been sizable.
200. *Bar Harbor Times*, Saturday, September 22, 1917. From the Archives of Dunsink Observatory, Dublin, Ireland.
201. *Princeton Herald*, February 28, 1930. From the Archives of Dunsink Observatory, Dublin, Ireland.

202. Katherine Brünnow married Louis F. Faulkner of New York, son of Frederick W. Faulkner, an author.
203. Marguerite Brünnow married Nelson Titus Hoadley, son of Judge Carleton E. Hoadley of Washington.
204. *Princeton Herald,* February 28, 1930. From the Archives of Dunsink Observatory, Dublin, Ireland.
205. *Princeton Packet,* December 30, 1922. From the Archives of Dunsink Observatory, Dublin, Ireland.

Chapter 7. Jasper F. Cropsey: Hudson River School Painter

206. These paintings by Cropsey hang in the Bentley Historical Library. Three other works by Cropsey are in the University of Michigan's Museum of Art, including two sketches and a painting called "Autumn Hastings-on-Hudson" that was completed in 1886.
207. Talbot, William S., *Jasper F. Cropsey 1823-1900,* Garland Publishing, Inc., New York, 1977, p. 381. According to Cropsey's account book, he was paid $50 for his paintings (p. 314). The University's records indicate $25 was appropriated for this purpose.
208. C. W. Field was, in the 1880s, a neighbor of Professor Bull (see Chapter 8) in Gramercy Park, New York City.
209. Talbot, William S., *Jasper F. Cropsey 1823-1900,* 1977, p. 316, 318, 320.
210. Newington-Cropsey Foundation, letter from Henry P. Tappan to Maria Cropsey, April 21, 1856.
211. Correspondence between Tappan and the Cropseys is held in the archives of the Bentley Historical Library and the Newington-Cropsey Foundation.
212. Newington-Cropsey Foundation, letter from Henry P. Tappan to Jasper F. Cropsey, January 28, 1856; HPT to Maria Cropsey, April 21, 1856.
213. New York Historical Society, *Jasper F. Cropsey: Artist and Architect,* New York, 1987, p. 163.
214. Ibid., p. 136.
215. See Samuel Taylor Coleridge's poem about Genevieve entitled *"Love,"* which is one of four Coleridge poems included in William Wordsworth's *Lyrical Ballads, and Other Poems 1797-1800,* James Butler and Karen Green, eds., Cornell University Press, 1992, Appendix IV. See the complete text of *Love* in Appendix 5.
216. Perry, Charles M., *Henry Philip Tappan,* 1933, p. 69.
217. Newington-Cropsey Foundation, letter from Henry P. Tappan to Jasper F. Cropsey, March 10-11, 1856; Henry P. Tappan to Jasper F. Cropsey, March 12, 1856. For the painting "Genevieve," Tappan paid $70 according to his letter, or $120 according to Cropsey's ledger.
218. Ibid., Henry P. Tappan to Jasper F. Cropsey, March 12, 1856.
219. The Cropseys had five daughters: Minnie (1850), Jennie (1852), Rose (1858-1892), Lilly (1859-1889), and Constance (1879).
220. Newington-Cropsey Foundation, Henry P. Tappan to Maria Cropsey, April 1, 1856.

221. Ibid., Henry P. Tappan to Jasper F. Cropsey, March 28, 1856; Henry P. Tappan to Jasper F. Cropsey, July 28, 1856.
222. Talbot, William S., *Jasper F. Cropsey 1823-1900*, 1977, p. 381-382.
223. Some of Tappan's papers were later donated to the University by Rudolph Brünnow at the request of Charles M. Perry when he was writing Tappan's biography, but they were largely papers that post-dated Tappan's removal from Michigan. Other papers were sent to Ann Arbor by Tappan to support the defense he wrote after being dismissed from the University, but the papers were entrusted to his attorney rather than to the University; their disposition is not known.
224. University of Michigan, Bentley Historical Library, Lucy Chapin Scrapbook, undated, unidentified newspaper clipping

Chapter 8. The Observatory's Architect

225. University of Michigan, *Regents Proceedings*, 1837-1864, p. 536-537.
226. The Board of Regents minutes book is held in the archives of the Bentley Historical Library, University of Michigan.
227. The term architect is used loosely. In the 1850s, many individuals designed buildings, not just trained architects.
228. Jones, Theodore F. (ed.), *New York University, 1832:1932*, 1933, p. 307-308.
229. *New York Times*, February 3, 1892, obituary for Richard Harrison Bull.
230. *New York Daily Tribune*, February 3, 1892, obituary for Richard Harrison Bull.
231. Ibid. For further details regarding Bull's timeservice activities, refer to Ian R. Bartky's book in progress, *Selling the True Time*.
232. Klein, Carole, *Glories of Gramercy Park Area: A Walking Tour*, Con Edison, New York, 1991. Note that the building's original hydraulic elevators still function today. Installed by Otis Brothers & Co., the elevators are powered by water pressure, rising and descending as the operator pulls on the cable that controls the valves of the water tanks. In more recent years, the building has served as the residence of famous actors, including James Cagney and Margaret Hamilton (Wicked Witch of the West in the 1939 film *The Wizard of Oz*).
233. Curiously, the earliest known photograph of the Detroit Observatory (see cover or Figure 14.1), circa 1856-58, was taken by T. D. Tooker. Whether there is any connection between the photographer and Dr. Bull has not yet been determined.
234. *New York Times*, February 3, 1892, obituary for Richard Harrison Bull.
235. Various biographical sketches on Bull were consulted: *Universities and Their Sons*, p. 55; *Twentieth Century Biographical Dictionary of Notable Americans*, The Biographical Society, Boston, 1904; *The National Cyclopedia*, New York, 1898, p. 472.
236. Dudley Observatory Archives, to J. H. Armsby papers, Box 1, folder 23, letter from O. M. Mitchel to Executive Committee dated January 13, 1853.
237. George Roberts Perkins was a self-taught surveyor who taught mathematics in Utica, New York. He authored numerous mathematics textbooks between 1842-56—some with applicability to astronomy—and was awarded an honorary LL.D. degree in the mid-1850s. He lived in Albany during the construction of the Dudley Observatory, which he supervised.

238. Dudley Observatory Archives, to J. H. Armsby papers, Box 1, folder 23, letter from O. M. Mitchel to Executive Committee dated January 13, 1853.

239. *New York Daily Times*, "The Dudley Observatory," Monday, September 1, 1856 (found in the Dudley Observatory Archives); Kohlstedt, Sally G., *Formation of the American Scientific Community: The American Association for the Advancement of Science 1848-60*, University of Illinois Press, 1976, Appendix entry for Perkins.

240. *New York Daily Times*, "The Dudley Observatory," September 1, 1856 (found in the Dudley Observatory Archives).

241. Ibid.

242. The Hamilton College Observatory was subsequently named the Litchfield Observatory after E. C. Litchfield, a donor to the Detroit Observatory who later endowed the Hamilton College Observatory in order to keep it operating. The building is no longer extant, although the pier for the refracting telescope was left in place. The pier now stands alone between buildings as a monument to C. H. F. Peters, whose career was invested in the observatory.

243. *Harper's New Monthly Magazine*, "Astronomical Observatories of the United States," Part I, Vol. 13, June 1856, p. 50.

244. Hamilton College Library, C. H. F. Peters papers.

245. Ibid., "Specifications for building addition to Observatory." Dr. C. H. F. Peters was director of the Observatory.

Chapter 9. Construction of the Detroit Observatory

246. University of Michigan, *Regents Proceedings*, 1837-1864, p. 538.

247. University of Michigan, Bentley Historical Library, Henry N. Walker papers, letter from Henry N. Walker to Silas Douglass dated August 8, 1853.

248. It is interesting to note that the property deed, dated Nov. 18, 1853, states that Mrs. Benham was taken aside and privately asked whether she freely agreed to the sale "without any fear or compulsion from her said husband or any other person."

249. Samuel Pettibone was the Washtenaw County surveyor; he resided on the south side of Huron St. between Division and State. See Stephenson, O. W., *Ann Arbor: The First Hundred Years*, p. 452.

250. University of Michigan, *Catalogue of the Corporation, Officers and Students of Medicine, Arts and Sciences in the University of Michigan, 1852-53*, Detroit, 1853, p. 34.

251. Tappan, Henry P., *Review by Rev. Dr. H. P. Tappan*, 1864, p. 17.

252. University of Michigan, Bentley Historical Library, Observatory Records, Box 2, copies of the original handwritten deeds, p. 77-79.

253. University of Michigan, Bentley Historical Library, UM Regents Collection, Box 45, folder 1718, letter from Ford, Armor, Gunn and Palmer to Hon. Alvah Sweetser dated January 8, 1864.

254. *Daily Free Press*, August 12, 1853, Detroit, p. 1.

255. D. H. Sperry was a cabinet maker and furniture dealer in Ann Arbor—Manufacturer of Sofas, Lounges, Bureaus, Tables, Ottomans, Coffins, Bedsteads, etc. His business was located on Huron St. between Fourth and Fifth Ave.; his

residence was at the corner of Huron and State St. See Stephenson, O. W., *Ann Arbor: The First Hundred Years*, p. 443, 453.

256. C. S. Goodrich, Jr., dealer in stoves, tin and copper ware, had his store on the east side of the Public Square, northeast corner of Main and Ann; his residence was on the west side of Fifth between Ann and Catherine Streets. From Stephenson, O. W., *Ann Arbor: The First Hundred Years*, p. 443, 451.

257. The names of the subcontractors and the amounts paid are from the University of Michigan, *Regents Proceedings*, 1837-1864, December 1855, p. 630, 681-682. Multiply the amounts paid by about 15 to get approximate costs in 1998 dollars. The costs listed represent the portions of the work at the Observatory that can be documented but are not assumed to be inclusive.

258. *Frank Leslie's Illustrated Newspaper*, May 7, 1859, p. 363.

259. The mansion was named Grace Hill after Litchfield's wife, Grace Hill Hubbard, although A. J. Davis called it "Ridgewood." The building survives, although the exterior stucco has been removed. Re-dubbed Litchfield Villa, it has long been occupied by the New York City Parks Department. A. J. Davis also designed, in 1853, a residence in Detroit for Bela Hubbard, Grace Hill Hubbard Litchfield's brother, called Vinewood. Bela Hubbard, a Michigan scientist and explorer after whom Hubbard Lake is named, was also a donor to the Detroit Observatory.

260. Stephenson, O. W., *Ann Arbor: The First Hundred Years*, 1927, p. 434.

261. Chase, A. W., *Dr. Chase's Recipes; or Information for Everybody*, published by the author, Ann Arbor, 1864, p. 322-324.

262. Ibid., p. 324.

263. Quinn Evans/Architects, *The Detroit Observatory*, Final Report, October 1996, p. 31. This 30-year estimate is the conclusion of Quinn Evans/Architects of Ann Arbor, based on photographic evidence and paint analysis performed by historic paint specialist Frank Welsh.

264. Newton, Roger Hale, *Town & Davis: Architects*, Columbia University Press, New York, 1942, p. 302.

265. The Elmes Engineering Works boasts of hauling 42,000 pounds, but this is a gross exaggeration of what the pier weighed.

266. Jones, Bessie Zaban, *The Harvard College Observatory*, 1971, p. 52-53, 57.

267. University of Michigan, Bentley Historical Library, E. C. Walker papers, letter to E. C. Walker from President Erastus O. Haven dated November 16, 1865.

268. University of Michigan, *Regents Proceedings*, 1837-1864, December 1855, p. 597.

269. Ibid., p. 597.

270. This clock, another similar clock from a later date, two other Regulator clocks, and the Tiede clock, all used at the Observatory, are still extant.

271. Baldwin, Lelland D., *Pittsburgh: The Story of a City*, University of Pittsburgh Press, 1937, p. 221.

272. University of Michigan, *Regents Proceedings*, 1864-1869, September 1866, p. 164.

273. A clerestory is a windowed wall used for light and ventilation. At the Detroit Observatory, the clerestory also provides head room at the top of the stairs into the dome room (see Figure 9.7 and others).

274. The Buhls were among the original donors to the Detroit Observatory.

275. Fence posts were provided by Jas. Black for $43.75, and D. E. Wines constructed the fence for $150 (see *Regents Proceedings*, 1837-1864, p. 681-682).

276. University of Michigan, *Regents Proceedings*, 1837-1864, December 1855, p. 601, 631, 683.

277. This refers to Huron Street, which still serves as a major traffic artery.

278. University of Michigan, *Regents Proceedings*, 1837-1864, December 1855, p. 575.

279. University of Michigan, Bentley Historical Library, E. C. Walker papers, letter to E. C. Walker from President Erastus O. Haven dated November 16, 1865.

280. The wooden form used to construct these archways still resides in the Observatory's basement, having been set aside by workers.

281. University of Michigan, *Regents Proceedings*, 1891-1896, September 1896, p. 648.

282. Only one of these original bookcases is extant. It was used in the 1997-98 restoration as a model to replicate the original, simulated wood grain finish that was used to dress up the bookcases and other woodwork by making it look like natural oak.

Chapter 10. Additions and Relocation Attempts

283. Tappan, Henry P., *Review by Rev. Dr. H. P. Tappan*, 1864, p. 17.

284. University of Michigan, Bentley Historical Library, letter from Franz Brünnow to Andrew D. White dated Feb. 5, 1866, reel 6.

285. Brünnow's observatory plans have not been located, and are presumed lost.

286. University of Michigan, *Regents Proceedings*, 1864-1869, August 1865, p. 104-105.

287. Ibid., 1864-1869, June 1868, p. 274.

288. Ibid., 1837-1864, September 12, 1860, p. 918.

289. Ibid., 1864-1868, August 1868, p. 349.

290. Ibid., 1880-1886, November 1883, p. 416.

291. University of Michigan, *Encyclopedic Survey*, Vol. IV, p. 1742.

292. University of Michigan, *Regents Proceedings*, 1876-1880, January 1878, p. 190.

293. The orbits of two planets, Venus and Mercury, lie between the sun and earth, occasionally crossing the face of the sun. This occurs frequently for Mercury and rarely for Venus. Observation of a transit from various places assisted the determination of the distance of the sun from earth. The rare transit of Venus last occurred in 1882, and will occur again in 2004.

294. University of Michigan, *Encyclopedic Survey*, Vol. IV, p. 1742.

295. University of Michigan, *Regents Proceedings*, 1876-1880, March 1880, p. 491.

296. Only the objective lens and cell survive from the refracting telescope. The transit telescope and chronometer survive, and are on display at the Detroit Observatory.

297. University of Michigan, *Michigan Alumnus*, June, 1903, p. 420-421.

298. Sketch by J. D. Telfer dated 1-21-66, from the files of the University Planner's office, University of Michigan.

299. University of Michigan, *Regents Proceedings*, 1975-1978, March 1976, p. 274.

Chapter 11. James Craig Watson

300. The biographical facts of Watson's life in this section are taken primarily from four sources: *The Sidereal Messenger*, "Biographical Memoir of James Craig

Watson," by Prof. George C. Comstock, Vol. VII, 1888, p. 273-286; Elliott, Clark A., *Biographical Dictionary of American Science*, p. 267-268; *Michigan University Book*, Richmond, Backus & Co., Detroit, 1888, p. 52; Beakes, Samuel W., *Past and Present of Washtenaw County Michigan*, S. J. Clarke Publishing Co., Chicago, 1906, p. 10-14. Where facts conflicted, the author selected the most credible.

301. University of Michigan, Bentley Historical Library, James C. Watson papers, Box 1, letter from B. A. Gould to James C. Watson dated September 16, 1868.
302. *Harper's New Monthly Magazine*, "The Zone of Planets Between Mars and Jupiter," Vol. 10, February 1855, p. 344.
303. University of Michigan, *Encyclopedic Survey*, Vol. II, p. 450
304. University of Michigan, Bentley Historical Library, Andrew D. White papers, letter from Lizzie to Aimèe dated July 4, 1861, reel 6. Hamilton College had no funds to pay Peters' salary during this time, so he was probably supplementing his income by working on the *Nautical Almanac*.
305. Hamilton College, C. H. F. Peters papers, unidentified newspaper clippings dated November 7, 1865 and January 19, 1866.
306. Data are from Burritt, Elijah H., *Geography of the Heavens*, New York, 1873, p. 231-232. See Appendix 2 for a complete list of Watson's asteroids.
307. The ecliptic is the apparent path of the sun among the stars; the intersection plane of the earth's solar orbit. (American Heritage Dictionary.)
308. University of Michigan, *Encyclopedic Survey*, Vol. II, p. 450.
309. University of Michigan, Bentley Historical Library, James C. Watson papers, Box 1, 2-page legal agreement signed by A. W. Chase dated August 1, 1873.
310. University of Michigan, Bentley Historical Library, James C. Watson papers, Box 1, letter from A. W. Chase to James C. Watson dated July 15, 1874. Note that the prescription included the ingredients camphor, laudanum, Jamaica ginger, and rye whiskey.
311. Chase, A. W., *Dr. Chase's Recipes; or, Information for Everybody*, published by the author, Ann Arbor, MI, 1864, p. xxvii.
312. The Van Vleck Observatory at Wesleyan University is named after John M. Van Vleck.
313. Pickering went on to become director of the observatory at Harvard College in 1877.
314. Henry Morton became the inaugural President of the Stevens Institute of Technology in Hoboken, New Jersey.
315. This photograph was probably taken by Prof. Henry Morton, Prof. Alfred Mayer, Prof. E. C. Pickering, or an assistant in their team of skilled photographers who were in Iowa representing the U.S. National Almanac Office to take photographs of the total solar eclipse. They took 42 perfect photographs that day, with exposures of 0.002 seconds each, which was an unusual feat in 1869. Five photographs were taken during the 83 seconds of totality. See Furman, Franklin D. (ed.), *Morton Memorial*, Stevens Institute of Technology, Hoboken, New Jersey, 1905, p. 173-174; 203. See Figure 1.4 for a photograph taken of the sun on the eclipse expedition to Burlington, Iowa in 1869.

316. U.S. Naval Observatory, *Reports of Observations of the Total Eclipse of the Sun, August 7, 1869*, Washington, D.C., 1869, p. 79.
317. Furman, Franklin D. (ed.), *Morton Memorial*, Stevens Institute of Technology, Hoboken, New Jersey, 1905, p. 173-174. Morton, who was Professor of Chemistry and Physics at the University of Pennsylvania, became President of the Stevens Institute of Technology in 1870.
318. Mitchell, Samuel A., *Eclipses of the Sun*, Columbia University Press, New York, 1935, p. 140.
319. U.S. Naval Observatory, *Reports of Observations of the Total Eclipse of the Sun, August 7, 1869*, Washington, D.C., 1869, p. 84.
320. Baum, Richard and Sheehan, William, *In Search of Planet Vulcan*, Plenum Press, New York, 1997, p. 196.
321. U.S. Naval Observatory, *Reports of the Solar Eclipses of July 29, 1878*, Washington, D.C., 1880, Plate 3, p. 123.
322. Ibid., p. 119.
323. Baum, Richard and Sheehan, William, *In Search of Planet Vulcan*, Plenum Press, New York, 1997.
324. Beakes, Samuel W., *Past and Present of Washtenaw County Michigan*, S. J. Clarke Publishing Co., Chicago, 1906, p. 10, 13-14.
325. Royal Astronomical Society, *Monthly Notices*, obituary, Vol. LII, London, p. 231.
326. Property described as Lots 13, 14, 15, and 16, south of the University House S. E. This would be located approximately where West Hall (formerly West Engineering building) currently stands. See University of Michigan, Bentley Historical Library, James C. Watson papers, Box 2, page in jotting book.
327. University of Michigan, Bentley Historical Library, Andrew D. White papers, letter from James C. Watson to A. D. White dated January 21, 1861.
328. Bishop, Morris, *Early Cornell 1865-1900*, Cornell University Press, 1962, p. 29, 41.
329. University of Michigan, Bentley Historical Library, Andrew D. White papers, letter from Henry S. Frieze to Andrew D. White dated Feb. 17, 1866, reel 6. Frieze's *Virgil's Aeneid* was published in 1860, with a second edition appearing in 1866.
330. "Lizzie" may be Miss Lizzie Gibson, a niece of Professor George P. Williams, who later become Mrs. E. E. DuBois.
331. The "Great Mogul" is most likely a reference to Tappan or Brünnow.
332. University of Michigan, Bentley Historical Library, Andrew D. White papers, letter from Lizzie to Aimèe dated July 4, 1861, reel 6.
333. University of Michigan, Bentley Historical Library, James C. Watson papers, Box 1, letter from James C. Watson dated September 12, 1862.
334. Tappan, Henry P., *Review by Rev. Dr. H. P. Tappan*, 1864, p. 31.
335. University of Michigan, Bentley Historical Library, James C. Watson papers, Box 1, letterbook pages 93-96, four letters dated April 3, 1860 from James C. Watson to various insurance agents.
336. Beakes, Samuel W., *Past and Present of Washtenaw County Michigan*, S. J. Clarke Publishing Co., Chicago, 1906, p. 10, 13-14.
337. University of Michigan, Bentley Historical Library, James C. Watson papers, Box 1, letter from James C. Watson to Lte. Gillis dated September 16, 1861.

338. Beakes, Samuel W., *Past and Present of Washtenaw County Michigan*, S. J. Clarke Publishing Co., Chicago, 1906, p. 31.
339. University of Michigan, Bentley Historical Library, James C. Watson papers, Box 1, letter from James Craig Watson to Andrew D. White dated Dec. 2, 1860, reel 6.
340. Ibid., letter from James Craig Watson to Andrew D. White dated July 1865, reel 5.
341. Ibid., letter from James Craig Watson to Andrew D. White dated July 1865, reel 5.
342. Ibid., letter from James Craig Watson to Andrew D. White dated Nov. 20, 1866, reel 7.
343. University of Michigan, *Regents Proceedings*, 1864-1869, June 1864, p. 49. For a list of 95 books owned by Watson c. 1860, see his jotting book, University of Michigan, Bentley Historical Library, papers of James C. Watson, Box 2.
344. Jones, Bessie Zaban, *The Harvard College Observatory*, 1971, p. 110-112.
345. University of Michigan, *The Chronicle*, Vol. V, September 1873 - June 1874, Ann Arbor, 1874, p. 199.
346. University of Michigan, *Encyclopedic Survey*, Vol. II, p. 453.
347. University of Michigan, Special Collections Library, Brünnow papers, letter from Franz Brünnow to Edward P. Evans dated July 28, 1865.
348. *The Sidereal Messenger*, "Biographical Memoir of James Craig Watson," by Prof. George C. Comstock, Vol. VII, 1888, p. 284.
349. University of Michigan, Bentley Historical Library, James C. Watson papers, jotting book, page numbered 62.
350. University of Michigan, Bentley Historical Library, James C. Watson papers, letter from H. Bruhns, University of Leipzig, to James C. Watson dated May 8, 1869.
351. University of Michigan, Bentley Historical Library, James C. Watson papers, letter from H. A. Newton, Yale University, to James C. Watson dated July 4, 1871.
352. University of Wisconsin Archives, George C. Comstock papers, letter from C. A. Young to G. C. Comstock dated August 22, 1887, as seen in Baum and Sheehan, *In Search of Planet Vulcan*, Plenum Press, New York, 1997, p. 291.
353. *American Journal of Science*, Third Series, Vol. XXI, Nos. 121-126, New Haven, 1881, p. 65.
354. Bless, Robert C., *Washburn Observatory 1878-1978*, University of Wisconsin, 1978, p. 2.
355. During a 1997 visit to the Washburn Observatory, the author found that the Solar Observatory is no longer extant.
356. Ibid., p. 2.
357. University of Michigan, Bentley Historical Library, Henry S. Frieze papers, letter from Henry S. Frieze to Andrew D. White from Ann Arbor dated December 9, 1880.
358. *American Journal of Science*, Third Series, Vol. XXI, Nos. 121-126, January to June 1881, New Haven, 1881, p. 64-65.
359. Ibid., letter from Andrew D. White in Berlin to Henry S. Frieze in Ann Arbor dated January 2, 1881.
360. This residence is no longer extant, nor is Twelfth Street, which ran south from South University Street between the present School of Education building and Lorch Hall.

Chapter 12. Brünnow Departs for the Dudley Observatory

361. Dudley Observatory Archives, Thomas W. Olcott papers, letter from O. M. Mitchel to Thomas W. Olcott dated May 2, 1859.
362. Dudley Observatory Archives, Trustees Minute Book, entry dated May 16, 1859.
363. Dudley Observatory Archives, Thomas W. Olcott papers, letter from O. M. Mitchel to Thomas W. Olcott dated May 5, 1859.
364. Harvard University, Houghton Library, Benjamin Peirce papers, letter from B. A. Gould to B. Peirce dated March 10, 1859.
365. Harvard University Archives, William C. Bond papers, letter from Franz Brünnow to George P. Bond dated May 13, 1859.
366. Ibid., letter from Franz Brünnow to George P. Bond from Ann Arbor dated June 2, 1859.
367. Tappan, Henry P., *Review by Rev. Dr. H. P. Tappan*, 1864, p. 30.
368. *Detroit Daily Tribune*, June 29, 1859. The director of the Cincinnati Observatory, Ormsby M. Mitchel, was also director, in absentia, of the Dudley Observatory in Albany.
369. Dudley Observatory, *The Dudley Observatory and the Scientific Council. Statement of the Trustees*, Albany, 1858, p. 163.
370. Harvard University Archives, Benjamin Peirce papers, letter from B. A. Gould to B. Peirce dated July 23, 1860.
371. Tappan, Henry P., *Review by Rev. Dr. H. P. Tappan*, 1864, p. 31.

Chapter 13. Tappan and Brünnow Leave Michigan

372. Ibid., p. 15.
373. See Perry, Charles M., *Henry Philip Tappan*, 1933, and Tappan, Henry P., *Review by Rev. Dr. H. P. Tappan*, 1864, for excellent accounts of the dismissal of Henry P. Tappan.
374. University of Michigan, *Regents Proceedings*, 1837-1864, p. 1054.
375. *Michigan Argus*, July 3, 1863, p. 2.
376. University of Michigan, *Michigan Alumnus*, August 1914, p. 585.
377. Ibid., p. 580.
378. *Michigan Argus*, January 1, 1864, p. 2.
379. Perry, Charles M., *Henry Philip Tappan*, 1933, p. 421-422.
380. Ibid., p. 419.
381. Ballif, Laurent, *Les Bains Public de Vevey: Cent ans de Bains Payes*, Éditions Cabédita, 1997, p. 190-197.
382. A copy in German of this funeral oration was donated to the University of Michigan Library in 1914 by Edward W. Pendleton, Class of 1872. The German was translated by Dr. W. Jordan and published in Frankfurt by H. Keller in 1865. Only a few copies of the address are still extant. An English translation is available at the New York State Archives, Albany, New York.
383. Perry, Charles M., *Henry Philip Tappan*, 1933, p. 271.
384. Ibid., p. 379.
385. Ibid., p. 383.

386. Henry S. Frieze, *A Memorial Discourse*, 1882, p. 7. Note that the copy of this document in the University Library bears the inscription in Julia Tappan's handwriting "Professor Evans from Mrs. Tappan."
387. Ibid., p. 13.
388. Ibid., p. 45.
389. University of Michigan, *Michigan Alumnus*, October 1912, p. 7, letter from Henry P. Tappan to Corydon L. Ford from Berlin dated Nov. 8, 1865.
390. Root, Margaret Cool, *Rackham Reports*, "Tappan, Bismarck, and the Bitter Connection: Reflections on Men and their Dogs in the Artful Memory," University of Michigan, 1986-1987, last page of text (pages not numbered).
391. University of Michigan, *Michigan Alumnus*, Vol. 1, No. 6, March 1895, p. 85-87; June 1895, p. 128-130; October 1901, p. 11-13; November 1901, p. 76-77; December 1901, p. 123-125.
392. Perry, Charles M., *Henry Philip Tappan*, 1933, p. 429.
393. Wayman, Patrick A., Dunsink Observatory, 1785-1985, Dublin, 1987, Appendix 16, notes from the diary of Dr. Piazzi Smyth.
394. Perry, Charles M., *Henry Philip Tappan*, 1933, p. 429. Note that the St. Bernard's name was "Porthos" (see *Michigan Alumnus*, August 1914, p. 578).
395. University of Michigan, Bentley Historical Library, Henry P. Tappan papers, folder 6, poem written by Rebecca Lloyd Tappan Brünnow, unidentified newspaper clipping.

Epilogue. Perpetuation of a Legacy:

396. University of Michigan, *Michigan Alumnus*, August 1914, p. 579.

Appendix 2. Scientific Publications of Franz F. E. Brünnow and James C. Watson while at the Detroit Observatory

397. Brünnow's publications in 1859-60, while he was Associate Director at the Dudley Observatory, for the purpose of this list, are considered to be done while at the Detroit Observatory because he maintained his directorship in absentia, and some of the research that led to papers published at Albany was performed in Ann Arbor. Other works were written, published, translated, or researched while at the Detroit Observatory.

Appendix 3. List of the Asteroids Discovered by James Craig Watson

398. *Sidereal Messenger*, Vol. VII, 1888, p. 286.

Appendix 3. Comets Discovered by John M. Schaeberle at the Detroit Observatory

399. Kronk, Gary W., *Comets: A Descriptive Catalog*, Enslow Publishers, Inc., New Jersey, 1984, p. 65, 67.

Appendix 5. Some Notable Astronomers Educated at the Detroit Observatory, 1854-1910

400. From Plotkin, Howard, "Henry Tappan, Franz Brünnow, and the Founding of the Ann Arbor School of Astronomers, 1852-1863," Annals of Science, vol. 37, 1980, p. 301-302; Chase, Theodore R., *Michigan University Book*, 1880; and other sources.

Bibliography

Abrahams, Peter, "Henry Fitz: A Preeminent 19th Century American Telescope Maker," *Journal of the Antique Telescope Society*, Summer 1994.

Adams, Charles Kendall, *Historical Sketch of the University of Michigan*, University of Michigan, Ann Arbor, 1876.

_____ (ed.), *Johnson's Universal Cyclopaedia*, Appleton, New York, 1895.

Albany Argus, "The Detroit Observatory of the University," 1858.

Albany Evening Journal, "Astronomers and Astronomical Instruments," August 30, 1856.

Alex, William, *Calvert Vaux: Architect & Planner*, Ink, Inc., New York, 1994.

Ambronn, L., *Handbuch der Astronomischen Instrumentenkunde*, Berlin, 1899.

Ambrose, Stephen E., *Undaunted Courage*, Simon and Schuster, New York, 1996.

American Association for the Advancement of Science 1848-60, University of Illinois Press, 1976.

American biographical history of eminent and self-made men, Michigan Volume, Cincinnati, 1878.

American Journal of Science, Third Series, Vol. XXI, Nos. 121-126, January to June 1881, New Haven, 1881.

Andrewes, William J. H. (ed.), *The Quest for Longitude*, Collection of Historical Scientific Instruments, Harvard University, 1998.

Andrews, A. D., *Irish Astronomical Journal*, "Cyclopaedia of Telescope Makers," Parts 3 and 4, Vols. 21 and 22, 1994 and 1995.

Annals of the Dudley Observatory, Volumes I and II, Argus Co. Printers, Albany, 1871.

Ann Arbor News, One Hundred Fifty Years of History, 1835-1985, Ann Arbor, 1985.

Arago, François, *Popular Astronomy*, 2 volumes, London, 1855.

Ashbrook, Joseph, *Astronomical Scrapbook*, Cambridge University Press, 1984.

Asmann, Edwin N., *The Telegraph and the Telephone: Their Development and Role in the Economic History of the United States, The First Century, 1844-1944*, Lake Forest College, 1980.

Baldwin, Lelland D., *Pittsburgh: The Story of a City*, University of Pittsburgh Press, 1937.

Ballif, Laurent, *Les Bains Public de Vevey: Cent ans de Bains Payes*, Éditions Cabédita, 1997.

Bar Harbor Times, Saturday, September 22, 1917.

Bartky, Ian R., *Selling the True Time* (in progress).

_____, *The Adoption of Standard Time*, Technology and Culture, Vol. 30, No. 1, January 1989, The University of Chicago, 1989.

Baum, Richard and Sheehan, William, *In Search of Planet Vulcan,* Plenum Press, New York, 1997.

Bedini, Silvio A., *Early Scientific Instruments and Their Makers,* Smithsonian Institution, Washington, D.C., 1964.

Beakes, Samuel W., *Past and Present of Washtenaw County Michigan,* S. J. Clarke Publishing Co., Chicago, 1906.

Berg, Herbert A., *Bela Hubbard, 1814-1898: A Biographical Sketch,* Michigan State University, 1967.

Bishop, Morris, *Early Cornell 1865-1900,* Cornell University Press, 1962.

Black, Mary, *Old New York in Early Photographs, 1853-1901,* New York Historical Society, Dover Publications, New York, 1973.

Bless, Robert C., *Washburn Observatory 1878-1978,* University of Wisconsin, 1978.

Blondheim, Menahem, *News Over the Wires: The Telegraph and the Flow of Public Information in America, 1844-1897,* Harvard University Press, 1994.

Boss, Benjamin, *History of the Dudley Observatory, 1825-1956,* Albany, 1968.

Bruce, Robert V., *Launching of Modern American Science,* 1846-1876, Alfred A. Knopf, NY, 1987.

Brünnow, Franz, *Astronomical Notices,* Volumes 1 (1858) through 29 (1862), Ann Arbor, and Albany (1859-60).

_____, *Lehrbuch der spärischen astronomie,* Berlin, 1851.

Burritt, Elijah H., *Geography of the Heavens,* New York, 1873.

Burt, John S., *They Left Their Mark: William Austin Burt and His Sons, Surveyors of the Public Domain,* Landmark Enterprises, Ranch Cordova, CA, 1986.

Burt, William A., *Description of the Solar Compass,* Geiger & Christian, Detroit, 1844.

Butkowsky, Harry A., *Astronomy and Astrophysics National Historic Landmark Theme Study,* National Park Service, May 1989.

Campbell, William W., *A Handbook of Practical Astronomy,* Inland Press, Ann Arbor, 1891.

Chamberlain, Joshua L. (ed.), *Universities and Their Sons,* R. Herndon Co., Boston, 1898.

Chase, A. W., *Dr. Chase's Recipes; or Information for Everybody,* Ann Arbor, 1864.

Chase, Theodore R., *Michigan University Book,* 1880.

Chauvenet, William, *Manual of Spherical and Practical Astronomy,* Lippincott, Philadelphia, 1871.

Cincinnati Observatory, *Inaugural Report of the Director,* June 30, 1868, Cincinnati, 1869.

_____, *Centenary of the Cincinnati Observatory,* November 5, 1943, Cincinnati, 1944.

Clark, Eliot, *History of the National Academy of Design,* Columbia University Press, New York, 1954.

Cooley, Thomas McIntyre, *Michigan: A History of Governments,* Houghton, Mifflin and Company, New York, 1897.

Daily Free Press, Detroit, August 12, 1853.

Description of the Observatory at Haverford College, Haverford College, 1857.

Detroit Free Press, December 28, 1853.

Donnely, Marian Card, *A Short History of Observatories,* University of Oregon Books, 1973.

Draper, John William, *Scientific Memoirs being Experimental Contributions to A Knowledge of Radiant Energy,* Arnor Press, New York, 1973.
_____, *An Address to the Alumni of the University of the City of New York at their Twenty-First Anniversary,* June 28, 1853, New York, 1853.
Dudley Observatory,
    Annals of the Dudley Observatory, Vol. I, Albany, 1866; Vol. II, 1871.
    Defence of Dr. Gould by the Scientific Council of the Dudley Observatory, Albany, 1858.
    Dudley Observatory and the Scientific Council. Statement of the Trustees, Albany, 1858.
    Inauguration of the Dudley Observatory at Albany, August 28, 1856, Albany, 1856.
    B. A. Gould, Jr., *Reply to the Statement of the Trustees of the Dudley Observatory,* Albany, 1859.
Dudley Observatory Archives,
    James H. Armsby Collection.
    Thomas W. Olcott Collection.
    Trustees Minute Book.

Erhardt, Roy, *Antique Clocks,* The House of Collectibles, New York, 1985.
Ehrmann, Emma S. G., "The University Observatories: From Sleepy Hollow to Portage Lake," *Quarterly Review,* 1949.
Elliott, Clark A., *Biographical Dictionary of American Science,* Greenwood Press, Westport, CT, 1979.

Farmer, Silas, *History of Detroit and Michigan,* Silas Farmer & Co., Detroit, 1884.
_____, *The Michigan Book: A State Cyclopedia,* Silas Farmer & Co., Detroit, 1901.
Farrand, Elizabeth M., *History of the University of Michigan,* Ann Arbor, 1885.
Fowler, Orson S., *A Home For All,* New York, 1850.
Four American Universities, Harper & Brothers, New York, 1895.
Frank Leslie's Illustrated Newspaper, May 7, 1859.
Frieze, Henry S., "Reminiscence of the Times of President Tappan," *Palladium,* Ann Arbor, 1888.
_____, *A Memorial Discourse on the Life and Services of Henry Philip Tappan, D.D., LL.D.,* University of Michigan, 1882.
Fuller, Edward G., *Michigan, a centennial history of the state and its people,* Lewis Publishing Co., Chicago, 1939, 5 volumes.
Furman, Franklin D. (ed.), *Morton Memorial,* Stevens Institute of Technology, Hoboken, New Jersey, 1905.

Geiger, Roger L., *To Advance Knowledge,* Oxford University Press, New York, 1986.
Gilliss, James M., *U.S. Naval Astronomical Expedition to the Southern Hemisphere,* 1849-50-51-52, Vol. III, "Observations to Determine the Solar Parallax," Washington, 1856.
Greene, Dascom, *Spherical and Practical Astronomy,* Boston, 1891.
Gurley, W. & L. E., *A Manual of the Principal Instruments used in American Engineering and Surveying,* Troy, New York, 1909.

Hackley, Charles W., *Elements of Trigonometry, plane and spherical,* Wiley & Putnam, New York, 1838.

_____, *Treatise on Algebra, containing the latest improvements,* Harper and Brothers, New York, 1846.

Hall, Asaph, Jr., *Determination of the Aberration Constant,* Lansing, MI, 1904.

Hamilton College Library, C. H. F. Peters papers.

Harper's New Monthly Magazine,
  "Application of Electro-magnetic Power to Railway Transit," Vol. 3, November 1851, p. 786-787.
  "Astronomical Observatories in the United States, Part I," Vol. 13, June 1856, p. 25-52.
  "Observatories in the United States," Part II, Vol. 48, March 1874, p. 518-39.
  "Can We Foretell the Weather?," Vol. 39, August 1869, p. 422-30.
  "The Coast Survey," Vol. 58, March 1879, p. 506-521.
  "The Electric Time Service," Vol. 56, October 1878, p. 665-671.
  "The Spots in the Sun," Vol. 40, May 1870, p. 818-824.
  "Some Talks of an Astronomer, Part I," Vol. 49, October 1874, p. 693-707.
  "Some Talks of an Astronomer, Part II," Vol. 49, November 1874, p. 825-841.
  "The First Century of the Republic: The Progress of the Exact Sciences," Vol. 52, December 1875, p. 82-100.
  "Seven Astronomers Royal," Vol. 49, November 1874, p. 894-896.
  "The American Railroad," Vol. 49, August 1874, p. 375-394.
  "The Zone of Planets Between Mars and Jupiter," Vol. 10, February 1855, p. 344-353.

Harrington, Mark W., *About the Weather,* Appleton, New York, 1901.

_____, *Weather making: ancient and modern,* Annual Report, Smithsonian Institution, 1894.

Harvard University Archives,
Benjamin Peirce papers.
William C. Bond papers.

Harvard University, Houghton Library, Benjamin Peirce papers.

Haslett, Charles, *The mechanic's, machinist's, & engineer's practical book of reference together with the Engineer's fieldbook* (ed. by Charles Hackley), Stringer and Townsend, New York, 1856.

Herman, Jan K., *A Hilltop in Foggy Bottom: Home of the Old Naval Observatory and the Navy Medical Department,* Department of the Navy, Washington, D.C., 1984.

Hinsdale, Burke A., *History of the University of Michigan,* Ann Arbor, 1906.

Hislop, Codman, *Eliphalet Nott,* Wesleyan University Press, 1971.

Hoyle, Fred, *Astronomy,* Doubleday, New York, 1962.

Hubbard, Bela, *Lake Superior Journal: Bela Hubbard's Account of the 1840 Houghton Expedition,* Northern Michigan University, Marquette, 1983.

_____, *Memorials of a Half-Century,* G. P. Putnam's Sons, New York, 1887.

_____, *Reports of William A. Burt and Bela Hubbard,* Detroit, 1846.

James, Mary Ann, *Elites in Conflict,* Rutgers University Press, 1987.

Jones, Bessie Zaban, *The Harvard College Observatory: The First Four Directorships, 1839-1919,* Belknap Press of Harvard University Press, 1971.

Jones, Theodore F. (ed.), *New York University, 1832:1932,* New York University Press, 1933.

_____, *New York Collection,* New York University, 1938.

Kestenbaum, Justin L. (ed.), *The Making of Michigan 1820-1860,* Wayne State University Press, Detroit, 1990.
King, Henry C., *The History of the Telescope,* Charles Griffin & Co. Ltd., London, 1955.
Klein, Carole, *Glories of Gramercy Park Area: A Walking Tour,* Con Edison, New York, 1991.
Kohlstedt, Sally G., *Formation of the American Scientific Community,* University of Illinois Press, 1976.
_____ and Rossiter, Margaret W., *Historical Writing on American Science,* Johns Hopkins University Press, Baltimore, 1985.
Kronk, Gary W., *Comets: A Descriptive Catalog,* Enslow Publishers, Inc., New Jersey, 1984.
Kuiper and Middlehurst (eds.), *Telescopes,* Chapter 6, "Transit Instruments," Chicago, 1960.

Lankton, Larry, *Cradle to Grave: Life, Work, and Death at the Lake Superior Copper Mines,* Oxford University Press, 1991.
Lerebours, Noel M., *A Treatise on Photography,* Arno Press, New York, 1973.
Lingenfelter, Paul E., *The Firing of Henry Philip Tappan, University Builder,* unpublished paper, 1970, University of Michigan, Bentley Historical Library, Henry Philip Tappan papers.
Litchfield Villa, League for the Restoration & Preservation of Litchfield Villa, Brooklyn, New York, 1980.
Loomis, Elias, *Recent Progress of Astronomy,* Harper & Brothers, New York, 1856.

Mabee, Carleton, *American Leonardo: A Life of Samuel F. B. Morse,* Alfred A. Knopf, New York, 1943.
Malone, Dumas (ed.), *Dictionary of American Biography,* Scribner's Sons, New York.
Marshall, Walter P., *Ezra Cornell (1807-1874): His Contributions to Western Union and to Cornell University,* Newcomen Society, New York, 1951.
Marwil, Jonathan L., *A History of Ann Arbor,* Ann Arbor Observer, Ann Arbor, 1987.
Mattison, Hiram, *A High School Astronomy,* New York, 1857.
May, George S., *Michigan: An Illustrated History of the Great Lakes State,* Windsor Publications, 1987.
McAlester, Virginia & Lee, *A Field Guide to American Houses,* Alfred A. Knopf, New York, 1992.
Mehlin, Theodore G., "Williams College Renovates Hopkins Observatory," a photocopied article given to the author by Professor Jay Pasachoff at Williams College.
Mennim, Eleanor, *Transit Circle: The Story of William Simms, 1793-1860,* 1992.
Michigan Argus, December 14, 1853; July 3, 1863; January 1, 1864.
Michigan Alumnus, November 1916.
Michigan Book, Inland Press, Ann Arbor, 1898.
Michigan History, "The Detroit Young Men's Society," June 1959.
Michigan Senate Document, No. 28, February 26, 1850.
Michigan University Book, Richmond, Backus & Co., Detroit, 1888.
Middleton, W. E. Knowles, *The History of the Barometer,* The Johns Hopkins Press, 1964.
Miller, Freeman D., "The Portage Lake Observatory of the University of Michigan," *Sky and Telescope,* Vol. IX, No. 4, February 1950.

Mitchell, Samuel A., *Eclipses of the Sun,* Columbia University Press, 1935.

Moore, Charles, *History of Michigan,* Lewis Publishing Company, Chicago, 1915, 4 volumes.

Morse, Edward Lind (ed.), *Samuel F. B. Morse: His Letters and Journals,* Houghton Mifflin, Boston, 1914.

Morse, Samuel F. B., *Examination of the Telegraphic Apparatus and the Processes in Telegraphy,* Washington, 1869.

Moyer, Claire Inch, *Silver Domes,* Big Mountain Press, Denver, 1955.

Multhauf, Robert P., *Holcomb, Fitz, and Peate: Three 19th Century American Telescope Makers,* United States National Bulletin 228, Smithsonian Institution, Washington, D.C., 1962.

Nagle, Paul C., *John Quincy Adams, A Public Life: A Private Life,* Alfred A. Knopf, New York, 1997.

National Archives, *List of Climatological Records in the National Archives,* Washington, 1942.

National Cyclopaedia of American Biography, New York, 1898.

Nature, "Observatories in the United States," July 9, 1874.

New Columbia Encyclopedia, Columbia University Press, New York, 1975.

New York City, Michelin Green Guide, Michelin Tire Corp., 1968.

New York Daily Times, "The Dudley Observatory," Monday, September 1, 1856.

New York Daily Tribune, February 3, 1892.

New York Historical Society, *Jasper F. Cropsey: Artist and Architect,* New York, 1987.

New York Times, February 3, 1892.

Newhall, Beaumont, *The Daguerreotype in America,* Duell, Sloan & Pearce, 1961.

Newington-Cropsey Foundation, Jasper F. Cropsey papers.

Newton, Roger Hale, *Town & Davis: Architects,* Columbia University Press, New York, 1942.

Nute, Grace Lee, *Lake Superior,* The American Lakes Series, Bobbs-Merrill Co., New York, 1944.

Peck, Amelia (ed.), *Alexander Jackson Davis: American Architect 1803-1892,* The Metropolitan Museum of Art, 1992.

Peckham, Howard H., *The Making of the University of Michigan 1817-1967,* University of Michigan Press, Ann Arbor, 1967.

Perry, Charles M., *Henry Philip Tappan: Philosopher and University President,* Ann Arbor, 1933.

Peters, Bernard C. (ed.), *Lake Superior Journal: Bela Hubbard's Account of the 1840 Houghton Expedition,* Northern Michigan University Press, Marquette, 1983.

Phillips, Sandra S., *Charmed Places: Hudson River Artists and Their Houses, Studios, and Vistas,* Harry N. Abrams, Inc., New York, 1988.

Pickering, Edward C., *Statement of the Work Done at the Harvard College Observatory During the Years 1877-1882,* Cambridge, 1882.

Plotkin, Howard, "Henry Tappan, Franz Brünnow, and the Founding of the Ann Arbor School of Astronomers, 1852-1863," *Annals of Science,* Vol. 37, 1980.

Princeton Herald, February 28, 1930.

Princeton Packet, December 30, 1922.

Quinn Evans / Architects, *The Detroit Observatory*, Final Report, October 1996.

Reade, Marjorie and Wineberg, Susan, *Historic Buildings, Ann Arbor, Michigan,* Ann Arbor Historical Foundation, 1992.

Rinhart, Floyd and Marion, *The American Daguerreotype,* University of Georgia Press, Athens, 1981.

Richmond, J. F., *New York and Its Institutions, 1609-1873,* New York, 1872.

Root, Margaret Cool, *Rackham Reports,* "Tappan, Bismarck, and the Bitter Connection: Reflections on Men and their Dogs in the Artful Memory," University of Michigan, 1986 1987

Royal Astronomical Society, *Monthly Notices,* obituary, Vol. LII, London, 1892,

Rudolph, Frederick, *The American College and University,* Vintage Books, New York, 1962.

Runes, Dagobert D., *Pictorial History of Philosophy,* Philosophical Library, New York, 1959.

Sagendorf, Kent, *Michigan, The Story of the University,* E. P. Dutton, New York, 1948.

Schaeberle, John M., *Selected Papers Read Before the Engineering Society of the University of Michigan,* "The Sun's Distance," Vol. 1, Ann Arbor, 1883, p. 33-38.

Scharf, Aaron, *An Album of Pictures and Words,* Harry N. Abrams, Inc., New York, 1975.

Scheffle, Joseph Viktor von, *Mountain Psalms,* translated from German by Franz Brünnow, T. Trubner, London, 1882.

Scientific American, "The Largest Telescope in the Country," Vol. 4, No. 14 new series, April 6, 1861.

Sidereal Messenger, Mitchel, O. M. (ed.), Cincinnati, Vol. I (1846), III (1884), VII (1888).

Smart, Charles E., *The Makers of Surveying Instruments in America Since 1700,* Regal Art Press, Troy, NY, 1962.

Smith, Shirley W., *Harry Burns Hutchins and the University of Michigan,* University of Michigan Press, Ann Arbor, 1951.

Smithsonian Institution Archives, National Museum of American History, Henry Fitz papers.

Smithsonian Institution, *Annual Report of the Board of Regents,* July 1894, Washington, 1896.

_____, *Miscellaneous Collections,* Vol. II, Washington, 1862.

Sobel, Dava, *Longitude,* Walker & Co., New York, 1995.

Spann, Edward K., *The New Metropolis: New York City, 1840-1857,* Columbia University Press, 1981.

Sperling, Norman, "Fair Play for Fitz: Henry Fitz Introduces the All-American Telescope," *Rittenhouse,* Journal of the American Scientific Instrument Enterprise, Vol. 3, No. 2, Issue 10, February 1989.

Spill, William A., *University of Michigan Beginnings,* Michigan Historical Commission, Lansing, 1929.

Stephenson, O. W., *Ann Arbor: The First Hundred Years,* Ann Arbor, 1927.

Stevens, Frank Walker, *New York Central Railroad,* Putnam's Sons, New York, 1926.

Stover, John F., *The Life and Decline of the American Railroad,* New York Oxford University Press, 1970.

Sturges, Mrs. Jonathan, *Reminiscences of a Long Life,* New York, 1894 (1996 reprint of an autobiography published privately in 1894).

Talbot, William S., *Jasper F. Cropsey 1823-1900*, Garland Publishing, Inc., New York, 1977.

Tappan, Henry P., Abraham Lincoln. Rede bei der gedächtnissfeir in der Dorotheenkircke zu Berlin, 2. mai, 1865, gehalten von dr. H. P. Tappan, corresp. mitglied der Französischen academie. Autorisirte übersetzung. Frankfurt am Main, 1865.

_____, *A Discourse, delivered by Henry P. Tappan, D.D. at Ann Arbor, Mich., on the occasion of his Inauguration as Chancellor of the University of Michigan, December 21st, 1852*, Advertiser Power Presses, Detroit, 1852.

_____, *Doctrine of the Will Determined by an Appeal to Consciousness*, Wiley and Putnam, New York, 1841.

_____, *Edwards's Inquiry into the Freedom of the Will*, New York, 1839.

_____, *Elements of Logic*, Wiley and Putnam, New York, 1844; Appleton, 1856.

_____, *The Growth of Cities*, R. Craighead, New York, 1855.

_____, *Progress of Educational Development*, Ann Arbor, 1855.

_____, *Public Education*, Detroit, 1857.

_____, *Review by Rev. Dr. H. P. Tappan of His Connection with the University of Michigan*, Free Press Steam Power Press Printing House, Detroit, 1864.

_____, *A Step from the New World to the Old, and Back Again; with Thoughts on the Good and Evil in Both*, Appleton, New York, 1852.

_____, *Treatise on the Will*, Glasgow, 1857.

_____, *University Education*, G. P. Putnam, New York, 1851.

Thompson, Robert L., *Wiring a Continent: The History of The Telegraph Industry in the United States 1832-1866*, Princeton University Press, 1947.

Thompson, Slason, *A Short History of American Railways*, Appleton, New York, 1925.

Trombino, Don and Pazmino, John, "Henry Fitz: Locksmith, Photographer, and Telescope Maker," *Skylines*, 1977.

Trow's New York City Directory, 1871, 1877, 1881.

Turner, George W., *Antique Scientific Instruments*, Blandford Press, England, 1980.

Twentieth Century Biographical Dictionary of Notable Americans, The Biographical Society, Boston, 1904.

Union College, *Union Worthies*, No. 22, "Henry Philip Tappan, Class of 1825," Doran, Kenneth T. and Hatcher, Harlan H., Schenectady, New York, 1967.

U.S. Naval Observatory, *Sky With Ocean Joined*, Washington, D.C., 1983.

_____, *Reports of the Solar Eclipses of July 29, 1878*, Washington, D.C., 1880.

_____, *Report of the Commission on the Site for Naval Observatory*, Washington, D.C., 1879.

University of Chicago, *The Yerkes Observatory*, University of Chicago Press, 1914.

University of Michigan
Bentley Historical Library,
William D. Anderson papers.
Lucy Chapin papers.
Jasper F. Cropsey sketchbook.
Faculty portraits.
Henry S. Frieze papers.
Alexander C. McGraw papers.

Charles H. Palmer family papers.
Henry Philip Tappan papers.
Henry Nelson Walker papers.
Observatory Records.
James C. Watson papers.
Andrew D. White papers.
Alexander Winchell papers.
Catalogue of the Corporation, Officers and Students of Medicine, Arts and Sciences in the University of Michigan, 1852-53, Detroit, 1853.
Chronicle, Vol. V, September 1873 - June 1874, Ann Arbor, 1874.
Encyclopedic Survey, 4 volumes.
Michigan Alumnus, 1895, 1902, 1903, 1904, 1912, 1914.
Publications of the Astronomical Observatory of the University of Michigan, Vol. I, 1912.
President's Annual Report, 1871 through 1881.
Records of the University Planner's Office.
Regents Proceedings, volumes spanning 1837-1990.
Report of the Curator of the University Museum to the Board of Regents, July 1, 1903 to July 1, 1904, Ann Arbor, 1905.
Report of the Director of the Detroit Observatory, Ann Arbor, 1881.
Special Collections Library,
    Brünnow papers.
    Edward Payson Evans papers.
University Magazine, July 1869.
University of Wisconsin Archives, George C. Comstock papers.
Utley, Henry M., *Michigan Pioneer & Historical Collections*, Vol. V.
_____ and Cutcheon, Byron M., *Michigan As a Province, Territory and State, the Twenty-Sixth Member of the Federal Union*, Publishing Society of Michigan, 1906.

Waite, Diana S., *19th Century Tin Roofing*, New York State Historic Trust, 1971.
Warner, Deborah Jean and Ariail, Robert B., *Alvan Clark & Sons: Artists in Optics*, second edition, Willman-Bell, Inc., Richmond, VA, 1995.
Warner, Deborah Jean, "Elisa Robinson and Henry Fitz," *Rittenhouse*, Journal of the American Scientific Instrument Enterprise, Vol. 1, No. 1, Issue 1, November 1986.
Watson, James Craig, *Theoretical Astronomy*, J. B. Lippincott, Philadelphia, 1868.
_____, *American watches: an extract from the Report on Horology at the International Exhibition at Philadelphia*, 1876, Robbins and Appleton, New York, 1877.
_____, *A Popular Treatise on Comets*, James Challen & Son, Philadelphia, 1861.
_____, *Tables for the calculation of simple or compound interest and discount and averaging of accounts*, Ann Arbor, 1878.
Wayman, Patrick A., *Dunsink Observatory 1785-1985: A Bicentennial History*, Royal Dublin Society, 1987.
Whitesell, Patricia S., *The University Facilities Crisis of the 1990s: A Causal Analysis Based on Historical and Economic Trends and a Case Study of 13 Major Research Universities* (Ph.D. dissertation, University of Michigan, 1994).
Whitnah, Donald R., *A History of the United States Weather Bureau*, University of Illinois Press, 1961.
Williams, Edwin, *New York As It Is in 1834*, J. Disturnell, New York, 1834.

Williams, Henry Smith, *The Great Astronomers,* Simon and Schuster, New York, 1930.
Wilson, James Grant (ed.), *Memorial History of the City of New-York,* Vol. IV, New York History Company, 1893.
Wilson, W. Hasell, *Railroad History,* Allen, Lane & Scott, Philadelphia, 1895.
Wordsworth, William, *Lyrical Ballads,* and Other Poems 1797-1800, James Butler and Karen Green, eds., Cornell University Press, 1992.

Zahner, Donald D., *Popular Astronomy,* "Henry Fitz, Telescope Maker," Vol. LVIII, No. 529, August/September 1964.
Zeller, Bob, *The Civil War in Depth: History in 3-D,* Chronicle Books, San Francisco, 1997.

Index